Maharishi Dayanand Saraswati Book Series- 10

The Vedāṅga Jyotiṣa

(Sanskrit Text, Roman Transliteration and Scientific English Translation)

by

Prof. Ravi Prakash Arya

Maharshi Dayanand Saraswati Chair (UGC)
Maharshi Dayanand University, Rohtak

AMAZON BOOKS, USA

in association with

INDIAN FOUNDATION FOR VEDIC SCIENCE

1051, Sector-1, Rohtak-124001, Haryana, India
Ph. 9650183260; 9313033917
vedicscience@rediffmail.com; vedicscience@gmail.com
website: https://vedic-sciences.com or
https://raviprakasharya.com

First Edition

Vikram Era: 2079 (c. 2022)
Kali Era: 5124
Kalpa Era: 1972949124
Brahma Era: 155521972949124

ISBN No. 978-93-94724-04-4

Printed by

Indian Foundation for Vedic Science, 1051, Sector-1,
Rohtak-124001, Haryana

Contents

Preface

Creation

The creation is the vāk/Veda (expression of the blue print of Brahman). Since this creation is exclusive work of Brahman, so it is verily said, एको वेद: - *eko vedaḥ* (the Veda is one). This vāk (expression of the blue print) of Brahman exists in two forms:

1. In the visible form as a universe.

2. In the audible form as vibrations or quanta.

Everybody sees and hears this vāk (creation), but very few except Brahmā are capable of reading and comprehending it. Everybody is not able to hear this vāk. That is why in the Ṛgveda (10.71.4), it is said:

उत त्व: पश्यन्न ददर्श वाचमुत त्व: श्रृण्वन्न श्रृणोत्येनाम् ।
उतो त्वस्मै तन्वं वि सस्ने जायेव पत्य उशती सुवासा: ॥

uta tvaḥ paśyanna dadarśa vācham
uta tvaḥ śṛṇvanna śṛṇotyenām
uto tvasmai tanvaṁ visasre
jāyeva patya uśatī suvāsā

There are some who on seeing the vāk (expression of the blue print of Brahman) in the form of creation does not see (understand) it, while another hearing the vāk (vibrations/quanta of the universe) does not hear it. The Veda unfolds itself to the deserving seer like the wife to her husband.

In other words, it can be said that creation is the expression of the vāk (Veda) and the vāk of Brahman is the blue print of creation. Thus both are mutually inter-dependant. The 'Ṛta' (the law of dynamism) remains the same at all levels of nature and the universe, while their expression as we know is quite varied. The underlying laws of Ṛta applied to Satya (eternal prakṛti) are the same even when they express themselves in different forms, shapes, components, and levels of sophistication, complexity, and completeness. The Veda has expressly disclosed this fact, as

एकं सद् विप्रा बहुधा वदन्ति।

ekam sad viprā bhudhā vadanti

There is only one underlying law or truth expressed in different forms due to the law of dynamism.

In its most abstract form, the law of Ṛta (dynamism) is nothing but the 'will of Brahman' that moves or activates the satya (prakṛti in its unified state /or inactive state) to undergo vikṛti or transform into visible state/active state. Human life and the world are also governed by it. This 'Will of Brahman', in its true sense, can also be appreciated through mathematical formulae, laws of physics, chemistry, biology, anatomy, or physiology, but can also be understood in terms of family, social and national structures. The manifestation of unified law in various forms takes place due to the factor of māyā. इन्द्रो मायाभिः पुरुरूप ईयते। If one is able to transcend this māyā factor in the universe, known variously as the 'unified law' or the 'will of Brahman' or the 'Veda', it will unfold itself to him. This unified law is the subject of subjective experience and its manifestation into various forms of the universe is the subject of objective assessment. It exists in the objectively assessed physical universe as a 'unified field', but on the subjective level of experience, it unfolds itself as Brahman (pure consciousness, self-referral soul (pure Being), and ultimate truth. So the unified field and unified law are the two sides of the same coin. In the context of subjective internal experience, it is the 'unified law', but in the context of the objective external universe, it is a 'unified field' in its abstract form.

Vedic visionaries/seers who were the masters of vision (*ṛṣyo mantradraṣṭāraḥ*) visualised the unified law of nature upto the minutest details beyond the limits of time and space in Samādhi. That unified law is called Veda.

The Vedas

The Vedas, as already informed, are the blueprint of the creation of the living and non-living world. This blueprint existed in the cosmos since creation. It was perceived directly by the high-spirited yogī-s coded in the form of the Vedas. The process of creation of living and the non-living beings is known as first operation/Yajña, whereas the documentation of the blueprint of this Yajña of creation is called the second operation/Yajña. When the first operation/ Yajña (creation) was over the Devas (High-spirited yogī-s) performed the second operation/Yajña (documentation of the laws governing the creation of the living and non-living world). This fact is confirmed by the Vedas as follows.

'The Divine beings like Brahmā and other seers

visualised the blueprint of creation (Yajña) that physically existed before them. This visualisation was the first Dharma (unified law of creation). These laws of creation speak highly of spiritual space (Brahman) where reside the divine beings and siddhas during mokṣa[1].'

In fact, all laws of creation visualised by the seers were coded in the forms of literary couplets or Chhandas. Those couplets or Chhandas were regarded as Dharmas, the laws governing the creation. This is why, Yāska (Nirukta, 1.20), an ancient Indian Vedic scholar alludes to the Origin of the Vedas (Chhandas) as:

'There were ṛṣis to whom was revealed Dharma (unified law governing creation coded in the form of the Veda Mantras or Chhandas). They taught this dharma to their followers who were devoid of it, through oral instructions. When the later generation of Ṛṣis was not able to retain this knowledge through oral instructions, all these laws were subjected to documentation just to aid them[2].

---

[1] यज्ञेन यज्ञमयजन्त देवाः तानि धर्माणि प्रथमान्यासन् ।
ते ह नाकं महिमानः सचन्त यत्र पूर्वे साध्या सन्ति देवाः ॥
yajñena yajñam-ayajanta
devās-tāni dharmāṇi-prathamānyāsana
te ha nākam mahimānaḥ sacanta
yatra pūrve sādhyāḥ santi devāḥ.

(The Ṛgveda, 1.164.50; 10.90.16; the Atharvaveda, 7.5.1; the *Vājasaneyī Saṁhitā*, 31.16; the *Taittirīya Saṁhitā*, 3.5.11.5; the *Taittirīya Āraṇyaka*, 3.2.7)

[2] *sākṣāt-kṛt-dharmāṇaḥ ṛṣayoḥ babhūvuḥ.*

The records of the Vedas also tell us that on the basis of first operation/yajña (creation) which occurred on account of oblations of all kinds of particles or forces, the origin of ṛcas contained in the Ṛgveda, sāmans contained in the Sāmaveda, yajuṣas contained in the Yajurveda and other Chhandas contained in the Atharvaveda took place in the cosmos. One of the seers of the Vājasaneyī SaŠhitā (31.7) sheds an ample good light on this fact as:

'From the Yajña (creation), in which oblations of all particles and forces were made, originated Ṛcas, Sāmans, Yajuṣas and other Chhandas (the Atharvaveda)[3].'

In fact, in that cosmic Yajña of creation, no material oblation was offered, as usual. However, ṛcas acted as the oblations of milk. Similarly, yajuṣas acted as oblations of Ghee, and the sāmans acted as the oblations of the Soma. This fact has been very carefully disclosed in the *Śatapatha Brāhmaṇa* (11.5.6.3,4,5). Here it may be pointed out that the sage to whom the knowledge of creation was revealed was Brahmā who passed this knowledge on to four ṛṣis, viz. Agni, Vāyu, Āditya, and Aṅgirā.

Classification of literary Couplets, or Chhandas into Ṛk, Yajuḥ, and Sāman:

As a result of the second great operation/Yajña (documentation of revealed knowledge in coded form) by the enlightened ṛṣi Brahmā, a huge number of couplets/Chhandas came into being dealing with mass-

---

3   *tasmād yajñātsarvahutaḥ ṛcaḥ sāmāni jajñire.*
     chandāṁsi jajñire tasmād yajus-tasmādajāyata.

energy existing at various levels of creation, viz. Bhumi (earth or observer space), dyau (the sun or light space), and field or Intervening space between Bhumi and Dyau. For example,

1. Agni as geothermal energy in the earth and as mass-energy of observer space.

2. Field energy or Vāyu in the space intervening earth and sun.

3. Sūrya as the light of the sun (star) and as the mass-energy of the light space.

Thus the knowledge received about the geothermal energy of earth or mass-energy of light space called (Agni) and its sub-forms was coded as ṛcā. The knowledge received about field energy (Vāyu), and electric force (Indra) and their forms existing in the space intervening earth and the sun was coded as yajuṣas and the knowledge of the sunrays and mass-energy of light space and its sub-forms was coded as sāmans. The Śvetāśvatara Upaniṣada (6.18), the Aitareya Brāhmaṇa (25.7) and the Manusmṛti (1.23) had it as:

'To make the second great operation/Yajña a success, three types of (brahmas) Vedic couplets were derived. From Agni (geothermal energy of the earth and mass-energy of observer space) were derived ṛcas; from Vāyu (field energy of space intervening earth and sun, or the observer space and the light space) were derived yajuṣas and from Sūrya (light of the sun or mass-energy of light space) were derived sāmans [4].

---

[4] *agni vāyu-ravibhyastu trayam brahma sanātanam*

N.B.: Here Brahma means 'Veda' that is why Brahmachārī is always known who undertakes the study of the Veda.

According to the Śatapatha Brāhmaṇa (11.5.8.3)

'On account of three forms of energy (tapta) in three spaces, three of the Vedas came into being. On account of Agni (geothermal energy of the earth and mass-energy of observer space) came into being; couplets called ṛcas or the Ṛgveda, on account of Vāyu (field energy or electric force of space intervening earth and sun; observer space and light space) came into being; yajuṣas or the Yajurveda and on account of Sūrya (light of the sun or mass-energy of light space) came into being the Sāmaveda or sāmans .'

The emergence of Saṁhitā literature

Later on, couplets dealing with ṛcas, yajusas, sāmans, and other chhandas were consolidated and compiled ṛṣi-wise and deity-wise into different literary styles and classified as the Ṛk Saṁhitā, the Yājuṣa-Saṁhitā, and the Sāma Saṁhitā respectively regardless of their earlier classification. During the course of the new classification, knowledge coded in metrical style was compiled as the Ṛk Saṁhitā (ṛgarcani). Knowledge coded in prose style was separately compiled in the name of the Yājuṣa Saṁhitā (*yat praśliṣṭaṁ paṭhitaṁ tat yajuḥ*) and the knowledge coded to the tune of cosmic vibrations was compiled under the caption the Sāma Saṁhitā. (*gitiṣu sāmākhyā*). The knowledge of miscellaneous nature coded metrically was compiled under the title Atharva

---

dudoha yajña siddyarathaṁ ṛg-yajuḥ-sāma-lakṣaṇam

Saṁhitā. So the Atharva Saṁhitā became the representative of all other three Saṁhitās.

Vedāṅgas

Vedāṅgas are six auxiliary disciplines of the Vedas that were developed in ancient times as the help book for the study of the Veda. There are six Vedāṅgas:

1. **Śikṣā (Phonetics)**: Śikṣā focused on the the accent, quantity, stress, melody and rules of euphonic combination of words of Vedic alphabets, so that the Vedic mantars may be recited properly.

2. **Chandas (Prosody)**: Chandas particularly deal with the system of composition of the Vedic mantars.

3. **Vyākaraṇa (Gramamr)**: Vyākaraṇa deals with rules of derivation of Vedic vocables involving their roots, suffixes and prefixes. In another words, it empowers a seeker to analyse Vedic words to explore their intended sense.

4. **Nirukta (Etymology)**: Etymology helps explain the cultural or scientific background of a Vedic vocabes, so that a proper intended sense of the archaic Vedic vocables may be derived.

5. **Kalpa (Figurative/ Imaginery Rituals)**: Kalpa texts have been devided into three parts:

(a) **Śrauta Sūtras**: The Vedas deal with the Yajña (process of creation). Śrauta Sūtras try to explain the various aspects of the process of creation by repeating them figuratively through imaginery yajñas. Their main focus is on standardizing procedures for figurative Śrauta yajñas developed to explain some or other aspect of the creation (b). **Gṛhya Sūtras**: Gṛhya Sūtras deals with the Sanskāras (processing of human beings).

Charaka defines the term Sanskāra as under:

संस्कारो हि गुणान्तरधानमुच्यते ।

samskāro hi guṇāntaradhānamuchyate.

[Meaning] Sanskara means inculcation of good values

Modern science has given stress upon the *sanskāra* of material things. Here there is a basic difference in the concept of Vedic Science and modern science. Vedic science gave more thrust on the sanskāra of human beings as compared to material things. According to the Vedic science, a human being who has not undergone through the certain process of *sanskāras* remains no better than an animal. As such a human being is not considered to be the actual social being. In material things also we can see, so long as they are in their crude form, they cannot have their practical application. For their practical application, the material things have to undergo a particular sanskāra or processing. For instance, an oil in its crude form cannot be used to operate cars, buses and airpalnes, but after undergoing the process of refinement it becomes usable. Similarly, a human being without inculcation of proper sanskāras cannot become a good social being. That is why the Gṛhya Sūtras were developed.

(c). **Śulva Sūtras**: For performing figurative Śrauta yajñas and Gṛhya yajñas yajña-altars of particular geometrical parameter are required. The Śulva sūtras deals with the method of construction of various types of altars.

6. **Jyotiṣa (Mathematical Astronomy)**: Jyotiṣa was developed as a technique of time calculation to know the timings of various figurative śrauta yajñas representing various aspects of creation and gṛhya yajñas.

Time Calculation

Techniques of time calculations were invented in this country in the Vedic period itself. This is proved by the following Ṛgvedic and other mantras:

द्वादशारं न हि तज्जराय वर्वति चक्रं परिद्यामृतस्य।
आ पुत्रा अग्रे मिथुनासो अत्र सप्त शतानि विंशति तस्थु: ।। ऋ.1.164.2

dvādaśāraṁ na hi tajjarāya varvati chakraṁ paridyāmṛtasya;
ā putrā agne mithunāso atra sapta śatāni viṁśati tasthuḥ.

[Meaning] The wheel of time (the Sun) having 12 spokes (rāśis) rotates in the space, but it does not wear out. O Agni, 720 pairs of sons (civil ahorātrās) ride this wheel.

द्वादश प्रधयश्चक्रमेकं त्रीणि नाभ्यानि क उ तच्चिकेत।
तस्मिन् त्साकं त्रिशता न शंकवोऽर्पिताः षष्टि न चलाचलास: ।। ऋ.1.164.48

dvādaśa pradhayaśchakramekaṁ trīṇi nābhyāni ka u
tachchiketa;
tasmin tsākaṁ triśatā na śaṁkavo'rpitāḥ ṣaṣṭhi na
chalāchalāsaḥ. Ṛg. 1.164.48

[Meaning] 12 spokes (months), one wheel (year), three navals (seasons); who understands these. These are caused by the rotation of the Earth about its axis and revolution around the Sun on its elliptic path of 360 śaṅkus (degrees) which do not get loosened.

त्रीणि च वै शतानि षष्टिश्च संवत्सरस्याहानि सप्त च वै शतानि च वै विंशतिश्च
संवत्सरस्याहोत्रयः। तै. ब्रा. 7.17
trīṇi cha vai śatāni ṣaṣṭiścha saṁvatsarasyāhāni sapta cha vai

śatāni cha vai viṁśatiścha saṁvatsarasyāhotrayaḥ. TBr. 7.17

[Meaning] A saṁvatsara has 360 days and 720 both days and nights.

वेदमासो धृतव्रतो द्वादश प्रजावतः ।
वेदा य उपजायते । ।

vedamāso dhṛtavrato dvādaśa prajāvataḥ ।
vedā ya upajāyate

[Meaning] Know the 12 months created by the law of nature and also so the 13th additional month.

From the aforementioned citations it is crystal clear that from the period of the *Ṛgveda* itself, Vedic people knew about the 12 zodiac signs, 12 months, 28 nakṣatras and the elliptic path of 360 degrees. They knew the civil days and tithis. They also knew the intercalary months (adhikamāsas) caused in the attempt to synchronisation the civil year, sidereal year and synodic year with the solar year. They also knew that the time (kāla) is the creation of the sun.

The *Atharvaveda* describes the Sun as kāla.

अश्वो कालो वहति ।

aśvo kālo vahati

In the upaniṣads, this concept has been explained further. Accordingly, the origin of time takes place after the Sun. Beyond the Sun, the kāla does not exist. There it is available in the form of akāla. Here, the need is to understand the concept of kāla and akāla. The word kāla is formed of the root √kal saṅkhyāne 'to count' and √kal gatau 'to move'. Yāska, the ancient Vedic scholar etymologizes kāla as

कालः कालयतेर्गतिकर्मणः ।

kālaḥ kālayatergatikarmaṇaḥ

All the above citations clearly explain that kāla is formed of the roots "meaning to count" and "to move". Firstly it signifies, that the concept of kāla (time) evolves with the movement of planets, satellites and stars in our universe. Secondly, kāla is associated with counting. As such, kāla in the Vedic sense means that which evolves due to the rotations/revolutions of heavenly bodies in our solar system and that which can be counted. Briefly, kāla means countable time and akāla means uncountable time. The origin of countable time takes place after the Sun. Beyond the sun, kāla remains uncountable.

The tradition of Jyotiṣa as a Vedāṅga text started from the Veda itself and we find Lagadha as the last proponent of the Vedāṅga tradition. Vedic people in the early age used to calculate time with nakṣatras as mile-stones, rather than rāśis (zodiac signs). This was because they used to calculate time on the basis of the actual observations of the positions of the heavenly bodies which is called the sāyana (tropical) system of calculation of time. The Aśvinī nakṣatra used to be called the first nakṣatra as the direction of the creation of planets and satellites in our solar system after the explosion in the solar nebula was found to be the direction of Aśvinī nakṣatra by the seers. So, Aśvinī nakṣatra was regarded as the first nakṣatra in the series. In the beginning of Kalpa (4,320,000,000 years ago) Indian part of the globe was also located in the southern hemisphere annexed with the landmass of Australia and Africa. The human beings were born or landed first in the history of human civilization on the earth on the Indian landmass located in the southern hemisphere of the globe, when first the northward movement of the Sun was observed, they also found heliacal rising of the Aśvinī nakṣatra just before sunrise. So, the year was started at the winter solstice by the actual observations of the beginning of the Uttarāyana, the northward motion of the sun with Aśvinī nakṣatra. The arrival of Aśvinīkumaras of Nāstyās in the early morning is described in several hymns of the Ṛgveda, such as 1.46.7; 157.1; 1.83.6; 1.180.1; 2.39.2; 4.45.2; 5.76.1; 5.77.1-2.

To quote a few of the mantras:

प्रातर्यावाणा प्रथमा यजध्वं पुरा गृध्रादररुषः पिबातः ।

प्रातर्हि यज्ञमश्विना दधाते प्रशसन्ति कवयः पूर्वभाजः । ।

prātaryāvāṇā prathamā yajadhvaṃ purā gṛdhrādararuṣaḥ pibātaḥ;

prātarhi yajñamaśvinā dadhāte praśasanti kavayaḥ pūrvabhājaḥ.

[Meaning] First give oblations for those that come in the morning. Let you drink soma (the nectarous air in the morning time) before the misers. The sages of old extol that Aśvinīs claim the oblations of yajña at the day break.

Thus the tradition of Vedāṅga Jyotiṣa also sets off with these observations. But the position of the heliacal rising of nakṣatras during Uttarāyana continues to change due to the precession of the Earth. So, modifications in the Vedāṅga Jyotiṣa were also carried out from time to time. When the last tradition of the current Vedāṅga Jyotiṣa came into being which was finally handled by Maharṣi Lagadha, Uttarāyana used to coincide with the heliacal rising of the Dhaniṣṭhā just before the sun rise. So, the Dhaniṣṭhā was taken as the first nakṣatra for the sake of time reckoning and the Vedāṅga calendar was then modified by the seer Lagadha accordingly. Here it may be pointed out that Uttarāyaṇa is 23.5⁰ south of the Earth. This is also called the line of Capricorn today, but it was the line of Dhaniṣṭhā during the commencement of the extant Vedāṅga tradition. The celestial longitude of Dhaniṣṭhā was 270⁰ on the elliptic circle. As such the position of all nakṣatras vis-a-vis original/niryaṇa(sidereal) position during the Vedāṅga period was as under:

Sr. No.	Name of Nakṣatra	Vedāṅga Period position	Original or Sidereal/Niryaṇa position
1.	Dhaniṣṭhā	270-283.20	293.20-306.40
2	Śatabiṣak	283.20-296.40	306.40-320
3	P. Bhādrapada	296.40-310	320-333.20
4	U. Bhādrapada	310-323.20	333.20-346.40
5	Revati	323.20-336.40	346.40-360
6	Aśvanī	336.40-350	0-13.20
7	Bharaṇī	350-3.20	13.20-26.40
8	Rohiṇī	3.20-16.40	26.40-40
9	Kṛttikā	16.40-30	40-53.20
10	Mṛgaśirā	30-43.20	53.20-66.40

11	Ārdrā	43.20-56.40	66.40-80
12	Punarvasū	56.40-70	80-93.20
13	Puṣya	70-83.20	93.20-106.40
14	Āśleṣā	83.20-96.40	106.40-120
15	Maghā	94.40-110	120-133.20
16	P.Phalgunī	110-123.20	133.20-146.40
17	U.Phalgunī	123.20-136.40	146.40-160
18	Hasta	136.40-150	160-173.20
19	Chitrā	150-163.20	173.20-186.40
20	Svāti	163.20-176.40	186.40-200
21	Viśākhā	176.40-190	200-223.20
22	Anurādhā	190-203.20	223.20-236.40
23	Jyeṣṭhā	203.20-216.40	236.40-240
24	Mūla	216.40-230	240-253.20
25	Pūrvāṣāḍhā	230-243.20	253.20-266.40
26	Uttarāṣāḍhā	243.20-256.40	266.40-280
27	Śravaṇa	256.40-270	280-293.20

5-Year Yuga System:

The Vedāṅga Jyotiṣa tradition follows the 5-year yuga system. While deciding on this issue, seers had minute observations. To make the calculation system scientific and completely astronomical, the five-year Yuga system was adopted. The reasons are as follows-

1. As the very name 'yuga' denotes, it is a meeting point of more than one object. It was observed that the Sun and the Moon meet at the same nakṣatra with a difference of some degrees after 5 years on Amāvasyā, which nakṣatra they start from, as the solar sidereal year consists of 365.256366 days and the lunar sidereal month consists of 27.321667 days. If we divide 365.256366 by 27.321667, we get:

$$365.256366 \div 27.321667 = 13.36874378858362$$

Multiplying 13.36874378858362 by 5 we get

$$13.36874378858362 \times 5 = 66.84371894291809$$

This means that after 5 years of doing 66 rounds the Moon will come closer to the same point where it met the Sun.

Multiplying .84371894291809 again by 5 we get:

$$.84371894291809 \times 5 = 4.21859471459045$$ which means that the Moon will come to the same point with a difference of 4^0.

1. The above calculation shows that after five years the Sun and the Moon meet in the same nakṣatra with a difference of 4^0. As mentioned above, Dhaniṣṭhā Nakṣatra was taken as the first nakṣatra owing to its heliacal rising during that period.

2. Secondly, it was found that 1830 days in one Yuga represented a least common multiple (LCM) of solar, sāvana, synodic lunar and sidereal lunar months. For example,

2. 60 solar months of 30.5 days each, make a Yuga of 60 x 30.5= 1830 solar days.

3. 61 sāvana months of 30 days each, which amounted to 61 x 30= 1830 sāvana days in a Yuga, requiring one adhikamāsa in 5 years for being in unison with the solar months.

4. 62 synodic lunar months (tithimāsas) of 29.532 days each, which make a Yuga of 62 x 29.532=1830.984 or 1831 days, requiring two adhikamāsas in 5 years for being in unison with solar months.

4. 67 sidereal lunar months (nakṣatra months) of 27.32 days each, which amounted to 67 x 27.32=1830.44 days, requiring 7 adhikamāsās in 5 years for being in unison with solar months.

As such, the 5-year Yuga of 1830 days represented all the months.

The years of the 5-Year Yuga were called Saṁvatsaras.

The Saṁvatsaras of 5-Year Yuga were named as

1. Saṁvatsara
2. Parivatsara
3. Idāvatsara
4. Anuvatsara
5. Idvatsara

The adhikamāsa in the middle of the Yuga was called

'Malimlucha' and the adhikamāsa at the end of the yuga was called 'Sansarpa'.

Nakṣatras

The Vedāṅga Jyotiṣa used 27 nakṣatras instead of 28 nakṣatras identified by the Vedic seers around the celestial equator or ecliptic path of the Moon. The reason was that all 27 nakṣatras are located within a radius of 20^0 of the celestial equator, but the abhiji nakṣatra is located far away from the celestial equator, i.e. around 62^0. So it was dropped from the list for the sake of time calculation. Another reason was mathematical and practical convenience, as the sidereal period of the Moon was found to be more close to 27 days than 28. The Vedic names of nakṣatras have also been listed (in ślokas 32-35) starting from the kṛttikādi list of Yajurveda.

Parvas/Pakṣas

In the Vedāṅga Jyotiṣa the 5-Year Yuga system was divided into 124 parvas / pakṣas of a synodic cycle. Since the Yuga started with Māgha Śukla Pratipadā on the day of Uttarāyaṇa when the Sun and the Moon met first time in Dhaniṣṭhā Nakṣatra. As such, all odd parvas represented the Śukla Pakṣa and even parvas the Kṛṣṇa Pakṣa.

Months

The calculations of the months of the Vedāṅga Jyotiṣa are based upon seasons and not upon nakṣatras. This is proved by the sixth and seventh ślokas of Yājuṣa Vedāṅga which says that Uttarāyaṇa takes place in Dhaniṣṭhā Nakṣatra and synodic lunar month Māgha month and

Dakṣiṇāyana begins in Śrāvaṇa month at the midpoint of Āśleṣā Nakṣatra. This proves that Māgha etc. synodic lunar months are associated with ayanas (solstices) and viṣuvas (equinoxes) and not with the nakṣatras of the Sun or the Moon. Not only these sidereal months were also allowed to be associated with the ayanas and viṣuvas, that is why there was a provision for two intercalary months (adhika māsas) in a 5-year yuga to keep the sidereal year and synodic year in sync. We know that a sidereal year consists of 354 days and a synodic year consists of 366 days, so there is a difference of 12 days in a year. This difference of 12 days increases by 60 days or two months of 30 days each in 5 years. That is why the Vedāṅga Jyotiṣa makes a provision of two adhika māsas every third and firth year of the 5-year yuga system to keep both sidereal and synodic lunar years in sync and there is no mention of any kṣaya month. Even the Vedic mantras talk about the adhika māsas and not the kṣaya māsas.

In addition to the above, from the references of the Rāmāyaṇa, the Mahābhārata, Purāṇas, the Sūryasiddhānta and Siddhānta Śiromaṇī, it is crystal clear that we find mention of months associated with the ṛtus (seasons).

Year Cycles

The Vedāṅga Jyotiṣa deals with synodic and sidereal cycles. Accordingly, the 5-year yuga had 62 months and that means 62 x 29.5 = 1830 days. The 5-year Yuga contains 1860 tithis, i.e. 372 tithis per year in a Yuga. So, to bring the sidereal cycle in unison with the synodic cycle, a provision of two adhikamāsas- one at the middle (3rd year) and another at the end of the Yuga (5th year) was done.

Bhāṁśas (Unit of Angle)

The Sun travels (27 x 5=) 135 nakṣatras in (372 x 5=) 1860 tithis. As such, one nakṣatra would consist of 1860÷135 = 124/9 tithis. So, Lagadha defines

1 nakṣatra = 124 bhāṁśas

And the Sun moves through 9 bhāṁśas in one tithi. So, we get

1 tithi= 9 bhāṁśas

The angular distance travelled by the sun in a half lunation, i.e. from pakṣa to pakṣa is called parva. Therefore-

1 parva= 15 tithis

1 parva= 15 x 9 (bhāṁśas) = 135 bhāṁśas

Accordingly, 360^0 circumference of the circle contains 124 x 27= 3348 bhāṁśas.

Unit of Time

The following units of time have been used in the Vedāṅga Jyotiṣa:

1 akṣara= 0.231 seconds (Time required to pronounce one syllable)

5 akṣaras= 1 kāṣṭhā (1.15 seconds)

124 kāṣṭhās = 1 kalā (2.388059 minutes)

10.05 kalās = 1 nāḍikā (1/60th of day or 24 minutes)

2 nāḍikā= 1 muhurta (48 minutes)

1 nāḍikā= 1/60 of a day or 24 minutes

1 nāḍikā = 10+1/20 kalās (10.05 kalās)

30 muhurtas= 1 ahorātra (day or 24 hours)

50 palas= āḍhaka/maśaka

4 āḍhakas = 1 droṇa

4 āḍhakas -3 kuṭas/kuḍvas= 1 nāḍikā (24 minutes)

The Moon is used for measuring time. The moon moves through 1809 nakṣatras in a yuga of 1830 days. So, a nakṣatra would consist of 1830/1809=610/603 days. As such Lagadha defines-

1 solar day= 603 kalās

And the Moon takes 610 kalās to move through one nakṣatra.

Divasāṁśas

In order to synchronize 30 tithis in a month with 30 civil days in a month and 124 parvas in a yuga, Lagadha also divided the day into 30 muhūrtas and 124 divasāṁśas. Accordingly-

1 muhurta = 2 nāḍikās

1 nāḍikā= 24 minutes

Again there are 1860 tithis in 1830 days = (1860/30) 62 tithis in (1830/30) 61 days, we have:

1 civil day= 124 divaśāṁśas

1 tithi = 122 divasāṁśas

Also, the term pāda is used for both quarter of nakṣatra, i.e. 124÷4=31 and quarter of the day, i.e. 124÷4=31.

The ratio of Longest and shortest day

In terms of the above units, the difference between the longest day on the summer solstice (Dakṣiṇāyana) and the shortest day of the winter solstice (Uttarāyaṇa) is said to be 6 Muhurtas. As the total variation occurs in half a year of 183 days, the daily variation would be 6/183=2/61 muhurtas per day. So, the length of the longest and the shortest day would be 18 and 12 muhurtas respectively, which have a ratio of 3:2.

Ayanas

The Moon moves through 1809 nakṣatras in the 10 ayanas of the yuga. So, it moves through about 1809÷10=181 nakṣatras in one ayana. This amounts to 6 sidereal revolutions and 19 nakṣatras. This would make a total 1810 nakṣatras in a yuga, so we have to reduce one nakṣatra at some stage. Lagadha does so at the end of the first ayana or 12th parva. At the end of the first ayana (12th parva) he drops Puṣya after Punarvasū and uses Āśleṣā, thus making 1809 nakṣatras

The 10 ayanas (solstices) of the five-year yuga cycle begin with one of the following nakṣatras:

1. First ayana (1st day) with Vasu (Dhaniṣṭhā nakṣtra)
2. Second (184th day) with Tvaṣṭā (Chitrā nakṣtra)
3. Third (367th day) with Bhava (Ārdrā nakṣtra)
4. Fourth (550th day) with Aja (Pūrvabhādrapada nakṣtra)
5. Fifth (733rd day) with Mitra (Anurādhā nakṣtra)
6. Sixth (916th day) with Sarpa (Āśleṣā nakṣtra)
7. Seventh (1099th day) with Aśvinau (Aśvinī nakṣtra)
8. Eighth (1282nd day) with Jala (Pūrvāṣāḍhā nakṣtra)

9. Ninth (1465ᵗʰ day) with Dhātā (Uttaraphālgunī nakṣtra)

10. Tenth (1648ᵗʰ day) with Ka (Rohiṇī nakṣtra)

Viṣuva

According to the Vedāṅga Jyotiṣa, viṣuva is taken to be the equator or zero degrees longitude. So, viṣuva would be taken as the middle point of ayana. Hence, there are two viṣuvas: one at vasanta sampat (vernal equinox) and another at śarad sampat (autumnal equinox). The first viṣuva occurs 93 tithis, i.e. 6 parvas and 3 tithis. Thereafter the viṣuvas repeat after 186 tithis, i.e. 12 parvas and 6 tithis. So, according to Lagadha, the number of viṣuva (equinox) and multiplied by 2, subtracting 1 from the product and multiplying the balance by 6 give the number of elapsed parvas of the viṣuv. Half of this gives is the tithi of viṣuva (equinox).

Ṛtus (seasons)

There are six seasons in one samvatsara starting from Tapas (Śiśira) at the winter solstice. They are:

1. Śiśira starts on Māgha S-1
2. Vasanta starts on Chaitra S-3
3. Griṣma starts on Jyeṣṭha S-5
4. Varṣā starts on Śrāvaṇa S-7
5. Śarad starts on Āśvina S-9
6. Hemanta starts on Mārgaśīrṣa S-11

Thereafter Śiśira ṛtu starts on Māgha S-13, Māgha K-10, Māgha S-7 and Māgha K-4 in successive years.

Thus each season occurs after 2 months and two tithis.

Solar and Lunar Nakṣatras

The Moon takes one day plus 7 kalās (i.e. 603+7=610 kalās) to move through one Nakṣatra. The Sun covers one Nakṣatra in 13 days and 9 parts of 5 days, i.e. 13+5/9 days, i.e. 13.5 days to cover one nakṣatra. Also, the Sun covers 135 (124+11) bhāṁśas, i.e. 1 nakṣatra + 11 bhāṁśas in one parva, so in 12 parvas it covers-

12 x (1 nakṣatra + 11 bhāṁśas)

or   12 nakṣatras + 132 bhāṁśas

or 13 nakṣatras + 8 bhāṁśas

As such, to find the solar nakṣatra at the end of a parva, divide the given parva by 12. Multiply the quotient by 8 and the remainder by 11 and add both. If it is Śukla pakṣa (bright half) and we want to know the bhāṁśas of the Moon, then add half of the bhāṁśa (total 124 parvas), i.e. 62. The lunar nakṣatras would be the same as solar nakṣatras at the end of even parvas when we have amāvasyā. However, at the end of the odd parvas, when we have pūrṇimā, the moon would have 62 more aṁśas, being 180 degrees (13.5 nakṣatras) away from the Sun.

Kṣaya Tithis

There are 1860 sidereal tithis in 1830 synodic lunar days, so we are to drop 30 tithis in a yuga to keep sidereal cycle and synodic cycle in sync. It is normally stated that one has to drop one tithi after 61 tithis, but the tradition of Vedāṅga Jyotiṣa tradition prescribes omitting tithis at the end of parva, making the parva end on the 29th day by reducing the 15th tithi.

One parva is equal to 1830/124 days, i.e. 14 days and

94 divasāṁśas. So the parva ends after 14 days and 94 divasāṁśas. This means about 2 divasāṁśa less for each tithi in a parva or about 31 divsāṁśas per parva. So, excess divasāṁśa has to be less than 31 for dropping a tithi. If a parva ends in the first pāda of a tithi, that tithi is dropped and the next tithi is started. This means a parva ending before mid-day must have 15th tithi abandoned. Thus the amāvasyās of the 14 even parvas and pūrṇimās of 16 odd parvas are to be kṣaya tithis. We see that a total of 30 kasaya tithis are required.

The Purpose of the Vedāṅga Jyotiṣa

Yajña has been a very significant part of Vedic culture. Yajña in fact is the process of creation and the Vedas are the knowledge of creation. All Śrauta yāgas represent one or another aspect of the creation symbolically or allegorically. That is why there is a Vedic dictum:

वेदा हि यज्ञार्थं प्रसिद्धाः
vedā hi yajñārtham prasiddhāḥ

Vedas were made public to understand the yajña (process of creation).

तस्माद्यज्ञात्सर्वहुतः ऋचः सामानि जज्ञिरे ।
तस्माद्यजुस्तस्मादजायत ।।
tasmādyajñātsarvahutaḥ ṛcaḥ sāmāni jajñire
tasmādyajustasmādajāyata

Agni (geothermal energy), Vāyu (air) and Sūrya (solar energy) are the main instruments behind the Yajña (the process of creation). The Ṛgveda, the Yajurveda and the Sāmaveda originated on account of these.

अग्निवायु रविभ्यस्तु त्रयं ब्रह्म सनातनम्

दुदोह यज्ञसिध्यर्थं ऋग्यजुसामलक्षणम् ॥ मनुस्मृति, 1.23

agnivāyu ravibhyastu trayaṁ brahma sanātanam;
dudoha yajñasidhyarthaṁ ṛgyajusāmalakṣaṇam. Manusmṛti,
1.23

The Ṛgveda, the Yajurveda and the Sāmaveda which
are the eternal source of the knowledge of Brahman are
respectively produced on account of agni (energy in
observer space/geothermal energy), Vāyu (air/field
energy) and Sūrya (solar energy). They are needed for
the success of Yajña (the process of creation).

In view of the aforementioned facts, the tradition of
performing Śrauta yāgās was made prevalent in this
country to understand the ādhidaivika (astronomical),
ādhyātmika (metaphysical) and ādhibhautika (physical
aspects of the process of the present creation. Since the
time factor was keenly involved in the performance of
these allegorical yāgās representing the various aspects of
the process of creation, so the tradition of the Vedāṅga
Jyotiṣa was developed. This helped the yājñikas to be
familiar with the timings of the parvas, months, ayanas,
viṣuvas, years and yugas, as specific timings were
prescribed for the performing specific yāgās.

Darśapūrṇamāsa iṣṭis (yajñas) need the knowledge of
the amāvasyās and pūrṇimās for their performance. Parva
days occur after every 14.75 days. The 12th śloka specifies
when a particular tithi has to be dropped, the iṣṭis could
be performed on the following pratipadā. Ṛtuyajñas were
performed on the days specified by śloka 11, i.e. Māgha
S-1, Chaitra S-3, Jyeṣṭha S-5, Śrāvaṇa S-7, Āśvina S-9 and

Mārgaśīrṣa S-11 in the first year and so on.

Dākṣāyaṇīya iṣṭis wer spilt into 2 parts-

1. Amānta months during 15 years with a kṣaya pakṣa at the end

2. Pūṇimānta months during the next 15 years with a kṣaya pakṣa at the end

Gavāmayana satras of one-year duration of 361 or 354 days starting from Uttarāyaṇa and ending in the Pravargya ceremony before starting of the next satra.

Agnichayana vidhi is lasting for 95 years as described in the Śatapatha Brāhmaṇa (Kāṇḍa 6-9).

Chāturmāsya yajñas to be performed according to the Śatapatha Brāhmaṇa (2.5) are-

1. Vaiśvadeva on phālguna Pūrṇimā near the beginning of Vasanta Ṛtu as Śiśira Ṛtu started on Māgha S-1.

2. Varuṇapraghāsa on Āṣāḍha Pūrṇimā near the beginning of Varṣā Ṛtu.

3. Śākamedha on Kārtika Pūrṇimā near the beginning of Hemanta Ṛtu.

4. Śunāśirīya in the last month of 3rd and 5th year.

Method of Ascertaining the Uttarāyaṇa/Dakṣiṇāyana

During the Vedāṅga period new-year was started with Uttarāyaṇa which shows that the year originally had a solar base. Now the question arises of how the first day of Uttarāyaṇa was ascertained during those days. The answer is very simple. The ancient Vedic scholars didn't have a wall clock to view its needle. They used to gaze at the needle of the clock of the universe. A Layman's observation may be somewhat defective but the gazing of a seer is very perfect. Had there been the approximation in gazing at stars' movements, the seers would not have coined the micro-units of time. They used to identify the parvas, months, years and yugas with the help of the movements of the concerned planets and stars. The Vedic time calculation system was purely astronomical. The seers were able to identify the movements of the stars and after composing texts on various sciences, they passed them on to the laymen in society. Their purpose was to present the scientific knowledge in a very simple manner, so that the people may grasp it and make good use of it.

The commencement of Uttarāyaṇa was also identified with a particular Nakṣatra. By the time of Maharṣi Lagadha, Uttarāyaṇa used to occur in Dhaniṣṭhā Nakṣatra, which means, Uttarāyaṇa used to coincide with the heliacal rising of the Dhaniṣṭhā just before sun rise. So, the Dhaniṣṭhā was taken as the first nakṣatra for the sake of time reckoning. When the heliacal rising of the Dhaniṣṭhā just before sun rise occurred, the Solar New-year used to be declared. Afterwords when the Sun and the Moon used to rise up in Dhaniṣṭhā nakṣatra on the following amāvasyā, the Luni-solar New year used to be announced.

Recensions of the Vedāṅga Jyotiṣa

There are two recensions of the Vedāṅga Jyotiṣa. One belongs to Ṛgveda known as Ārcha Jyotiṣa, another belongs to the Yajurveda known a Yājuṣa Jyotiṣa. The Ārcha Jyotiṣa has 35 verses and Yājuṣa Jyotiṣa has 45 verses. Both belong to a long uninterrupted tradition of the Vedāṅga Jyotiṣa. Although the contents of the both are the same. Here we have rendered translation of both the recensions with examples and explanations wherever required and necessary.

Epoch of Maharṣi Lagadha

Interestingly, the statement of the Vedāṅga Jyotiṣa (śloka 7) is quoted by Varāhamihira while pointing out the difference of time between him and the Vedāṅga Jyotiṣa which is helpful in ascertaining the epoch of the Vedāṅga Jyotiṣa and Varāhamihira also. Varāhamihira states:

आश्लेषार्द्धाद् दक्षिणमुत्तरमयनं रवेर्धनिष्ठाद्यम्।
नूनं कदाचिदासीद्येनोक्तं पूर्वशास्त्रेषु ॥ बृहत्संहिता, 3.1

āśleṣārddhād dakṣiṇamuttaramayanaṁ raverdhaniṣṭhādyam;
nūnaṁ kadāchidāsīdyenoktaṁ pūrvaśāstreṣu. Bṛhatsaṁtā, 3.1

The Sun's southern course began at one time from the latter half of Āśleṣā and the northern from the beginning of Dhaniṣṭhā. This must indeed have been the case, as it is so recorded in previous Śāstras (The Vedāṅga Jyotiṣa).

साम्प्रतमयनं सवितुः कर्कटाद्यं मृगादितश्चान्यत्।
उक्ताभावो विकृतिः प्रत्यक्षपरीक्षणैर्व्यक्तिः ॥ बृहत्संहिता, 3.2

sāmpratamayanaṁ savituḥ karkaṭādyam mṛgāditaśchānyat;
uktābhāvo vikṛtiḥ pratyakṣaparīkṣaṇairvyaktiḥ.

Bṛhatsaṁtā, 3.2

At present, the one course of the Sun commences from the beginning of Karkaṭaka (Cancer), and the other from the Makara. This is different from what has been stated above and can easily be ascertained by direct observations.

It is crystal clear that by the time of Varāhamihira, the Sun had receded its southward course from the mid-point of Āśleṣā (113⁰) of the Vedāṅga Jyotiṣa to the beginning of Karka / Cancer (90⁰). This makes an interval of 113⁰-90⁰= 23⁰ between the time of the Vedāṅga Jyotiṣa and Varāhamihira which is equal to 23 x 72 = 1656 years. It shows that Vedāṅga Jyotiṣa was written 1656 years before the period of Varāhamihira.

Similarly, at the time of Varāhamihira, the Sun had receded its northward course from the beginning of Dhaniṣṭhā (293⁰') of the Vedāṅga Jyotiṣa, to the beginning of Makara / Capricorn (270⁰). This makes an interval of 293⁰-270⁰= 23⁰ between the time of Vedāṅga Jyotiṣa and Varāhamihira which is equal to 23 x 72=1656 years. It also confirms that Vedāṅga Jyotiṣa was written 1656 years before the period of Varāhamihira.

Now the epoch of Varāhamihira can also be worked out on a similar pattern. Today the Sun had receded its northward course from the beginning of Makara / Capricorn (283⁰), of Varāhamihira's period, to Mula nakṣatra (245⁰). This makes an interval of 283⁰-246⁰= 37⁰ between the time of Varāhamihira and the modern epoch which is equal to 37x72=2664 years. It also confirms that Varāhamihira lived 2664 years ago or 2664-2019=645 BC.

Similarly, today the sun has receded its southward course from the beginning of Karka / Cancer (90°) of Varāhamihira's period to the Mṛgaśirā constellation (53°). This makes an interval of 90-53=37° between the time of Varāhamihira and the modern epoch which is equal to 37 x 72=2664 years. It shows also clearly that Varāhamihira lived 2664 years ago or 2664-2019=645 BC.

This makes it clear that Vedāṅga Jyotiṣa was written 2022+645+1656=4343 years ago. So the period of Vedāṅga Jyotiṣa is about 2301 BC.

याजुषज्योतिषम्

Yājuṣa-jyotiṣam

Introduction

अथ याजुषज्योतिषं प्रारभ्यते ।

atha yājuṣajyotiṣaṁ prārabhyate

Now we start Yājuṣa Jyotiṣa.

पञ्चसंवत्सरमयं युगाध्यक्षं प्रजापतिम् ।
दिनर्त्वयनमासाङ्गं प्रणम्य शिरसा शुचिः ॥1 ॥

pañchasaṁvatsaramayaṁ yugādhyakṣaṁ prajāpatim;
dinartvayanamāsāṅgaṁ praṇamya śirasā śuchiḥ.

Having saluted with bent head to Prajāpati, the governor of the five-year yuga cycle that involving days, seasons, ayanas and months;

ज्योतिषामयनं पुण्यं प्रवक्ष्याम्यनुपूर्वशः ।
सम्मतं ब्राह्मणेन्द्राणां यज्ञकालार्थसिद्धये ॥2 ॥

jyotiṣāmayanaṁ puṇyaṁ pravakṣayāmyanupūrvaśaḥ;
sammataṁ brāhmaṇendrāṇāṁ yajñakālārthasiddhaye.

I shall narrate in sequence the movement of heavenly bodies as accepted by the towering scholars of this field for the purpose of determining the time of yajñas.

Importance of Mathematical Astronomy

वेदा हि यज्ञार्थमभिप्रवृत्ताः कालानुपूर्व्या विहिताश्च यज्ञाः ।
तस्मादिदं कालविधानशास्त्रं यो ज्योतिषं वेद स वेद यज्ञान् ॥3 ॥

vedā hi yajñārthamabhipravṛttāḥ kālānupūrvyā vihitāścha yajñāḥ;
tasmādidaṁ kālavidhānaśāstraṁ yo jyotiṣaṁ veda sa veda

yajñān.

Vedas have indeed been revealed by the Brahman for the sake of understanding yajña (the process of creation). The process of creation involves the sequence of time. Therefore, the one who knows the Astronomy, the Śāstra of calculating time, understands the process of creation.

यथा शिखा मयूराणां नागानां मणयो यथा।

तद्वद् वेदाङ्गशास्त्राणां गणितं मूर्धनि स्थितम् ॥4॥

yathā śikhā mayūrāṇāṁ nāgānāṁ maṇayo yathā;
tadvad vedāṅgaśāstrāṇāṁ gaṇitaṁ mūrdhani sthitam.

Like the crest of peacock and the gems of cobras stand atop their head/hood, so does the mathematical astronomy stand atop all Vedāṅga Śāstras.

Nominal Definition of a Yuga

माघशुक्लप्रपन्नस्य पौषकृष्णसमापिनः।

युगस्य पञ्चवर्षस्य कालज्ञानं प्रचक्षते ॥5॥

māghaśuklaprapannasya pauṣakṛṣṇasamāpinaḥ;
yugasya pañchavarṣasya kālajñānaṁ prachakṣate.

Now the details will be given of five years yuga that starts with the bright fortnight of the synodic lunar month Māgha and ends with the dark fortnight of the synodic lunar month of Pauṣa. According to the Vedāṅga Jyotiṣa, the months are Amānta as per South Indian tradition. They start from Māgha Śukla Pratipadā and ends with Pauṣa Amāvasyā.

सौर संक्रान्ति	मास का नामकरण
मकर	माघ/ तप
कुम्भ	फाल्गुन/ तपस्य
मीन	चैत्र/मधु

मेष	वैशाख/माधव
वृषभ	ज्येष्ठ/शुक्र
मिथुन	अषाढ/शुचि
कर्क	श्रावण/नभ
सिंह	भाद्रपद/नभस्य
कन्या	आश्विन/इष
तुला	कार्तिक/ऊर्ज
वृश्चिक	मार्गशीर्ष/सह
धनु	पौष/सहस्य

Solar Saṁkrānti	Name of the Month
Makara	Māgha/ Tapa
Kumbha	Phālguna/Tapasya
Mīna	Chaitra/Madhu
Meṣa	Vaiśākha/Mādhava
Vṛṣabha	Jyeṣṭha/Śukra
Mithuna	Āṣāḍha/Śuchi
Karka	Śrāvaṇa/Nabha
Siṁha	bhādrapada/nabhasya
Kanyā	Āśvina/Iṣa
Tulā	Kārtika/Ūrja
Vṛśchika	Mārgaśīrṣa/Saha
Dhanu	Pauṣa/Sahasya

Commencement of five-year yuga

स्वराक्रमेते सोमार्कौ यदा साकं सवासवौ ।

स्यात् तदाऽऽदियुगं माघस् तपःशुक्लोऽयनं ह्युदक् ॥6॥

svarākramete somārkau yadā sākaṁ savāsavau;
syāt tadā"diyugaṁ māghas tapaḥśuklo'yanaṁ hyudak.

When the Sun and the Moon rise up in Dhaniṣṭhā Nakṣatra (on an Amāvasyā day), at that time, on the first day of bright half of synodic lunar month Māgha, solar seasonal month Tapas and Uttarāyana; commences the five year yuga-cycle.

The names of five Saṁvatsaras are as follows: Saṁvatsara, Parivatsara, Idāvatsara, Idvatsara and Vatsara.

The above statement shows that by the time of Yājuṣa Jyotiṣa, winter solstice was taking place in Dhaniṣṭhā Nakṣatra.

Note: One sidereal year consists of 365.25 days and one revolution of the Moon around the Earth with respect to a Nakṣatra takes place in 27.25 days. If we divide 365.25 by 27.25 we get 13.4036697248. This shows that the Moon and the Sun meet together on a year 13.4036697248 times. If we multiply the same by 5= we get 67.01. That means that the Moon and the Sun meet 67th time exactly in Dhanistha Nakstra (293.20^0-306.40^0) with a 1^0 difference in five years. Here it may be pointed out that the Sun and the Moon meet at the same point on the day of Amāvasyā.

प्रपद्येते श्रविष्ठादौ सूर्याचन्द्रमसावुदक् ।
सार्पार्धे दक्षिणार्कस् तु माघश्रावणयोः सदा ॥7॥
prapadyete śraviṣṭhādau sūryāchandramasāvudak;
sārpā'rdhe dakṣiṇārkas tu māghaśrāvaṇayoḥ sadā.

When situated at the beginning of Dhaniṣṭhā Nakṣatra, the Sun and the Moon begin to move the northward. When they reach the midpoint of Āśleṣā

Nakṣatra, the sun begins to move southward. Thus Uttarāyaṇa begins in Māgha month and Dakṣiṇāyana begins in Śrāvaṇa month.

Note: Interestingly, Varāhamihira quotes the above statement of Vedāṅga Jyotiṣa while pointing out the difference of time between him and Vedāṅga Jyotiṣa. Varāhamihira states:

आश्लेषार्द्धाद् दक्षिणमुत्तरमयनं रवेर्धनिष्ठाद्यम् ।
नूनं कदाचिदासीद्येनोक्तं पूर्वशास्त्रेषु ॥ बृहत्संहिता, 3.1

āśleṣārddhād dakṣiṇamuttaramayanaṁ raverdhaniṣṭhādyam;
nūnaṁ kadāchidāsīdyenoktaṁ pūrvaśāstreṣu. Bṛhatsaṁhitā, 3.1

The Sun's southern course began at one time from the latter half of Āśleṣā and the northern from the beginning of Dhaniṣṭhā. This must indeed have been the case, as it is so recorded in previous Śāstras (The Vedāṅga Jyotiṣa).

साम्प्रतमयनं सवितुः कर्कटाद्यं मृगादितश्चान्यत् ।
उक्ताभावो विकृतिः प्रत्यक्षपरीक्षणैर्व्यक्तिः ॥ बृहत्संहिता, 3.2

sāmpratamayanaṁ savituḥ karkaṭādyaṁ mṛgāditaśchānyat;
uktābhāvo vikṛtiḥ pratyakṣaparīkṣaṇairvyaktiḥ. Bṛhatsaṁhitā,3.1

At present, the one course of the Sun commences from the beginning of Karkaṭaka (Cancer), and the other from the Makara. This is different from what has been stated above and can easily be ascertained by direct observations.

So at the time of Varāhamihira, the Sun had receded its southward course from the mid-point of Āśleṣā (113⁰) of the Vedāṅga Jyotiṣa to the beginning of Karka/ Cancer (90⁰). This makes an interval of 113-90= 23⁰ between the time of Vedāṅga Jyotiṣa and Varāhamihira which is equal to 23 x 72=1656 years. It shows that Vedāṅga Jyotiṣa was written 1656 years before the period of Varāhamihira.

Similarly, at the time of Varāhamihira, the Sun had receded its northward course from the beginning of Dhaniṣṭhā (293⁰') of the Vedāṅga Jyotiṣa, to the beginning of Makara / Capricorn (270⁰). This makes an interval of 293⁰-270⁰= 23⁰ between the time of Vedāṅga Jyotiṣa and Varāhamihira which is equal to 23 x 72=1656 years. It also confirms that Vedāṅga Jyotiṣa was written 1656 years before the period of Varāhamihira.

Now the epoch of Varāhamihira can also be worked out on a similar pattern. Today the Sun had receded its northward course from the beginning of Makara/ Capricorn (283⁰), of Varāhamihira's period, to Mula nakṣatra (245⁰). This makes an interval of 283⁰-246⁰= 37⁰ between the time of Varāhamihira and the modern epoch which is equal to 37x72=2664 years. It also confirms that Varāhamihira lived 2664 years ago or 2664-2019=645 BC

Similarly, today the sun has receded its southward course from the beginning of Karka / Cancer (90⁰) of Varāhamihira's period to the Mṛgaśira constellation (53⁰). This makes an interval of 90-53=37⁰ between the time of Varāhamihira and the modern epoch which is equal to 37 x 72=2664 years. It shows that Varāhamihira lived 2664 years ago or 2664-2019=645 BC.

This makes it clear that Vedāṅga Jyotiṣa was written 2022+645+1656=4343 years ago. So the period of Vedāṅga Jyotiṣa is about 2301 BC.

घर्मवृद्धिरपां प्रस्थः क्षपाह्रास उदग्गतौ ।

दक्षिणेतौ विपर्यासः षणमुहूर्त्ययनेन तु ॥8 ॥

gharmavṛddhirapāṁ prasthaḥ kṣapāhrāsa udaggatau;
dakṣiṇetau viparyāsaḥ ṣaṇamuhūrtyayanena tu.

During the northward course of the sun (Uttarāyaṇa), the increase of day-time and decrease in night-time per day is equal to the time taken by one prastha[5] (sher or litre) of water to enter the Nāḍī-yantra (a tubular instrument made for this purpose). The vice-versa happens during the southward course of the Sun. There is a total increase or decrease of time equal to 6 muhurtas.

Note: One muhurta is equal to around 48 minutes. So, 6 muhurtas are equal to (48x6=288 minutes= 4 hours 48 minutes tentatively).

प्रथम सप्तमं चाहुरयनाद्यं त्रयोदशम् ।

चतुर्थं दशमं चैव द्विर्युग्मं बहुले त्वृतौ ॥9 ॥

prathama saptamaṁ chāhurayanādyaṁ trayodaśam;
chaturthaṁ daśamaṁ chaiva dviryugmaṁ bahule tvṛtau.

The first, seventh, and the thirteenth tithis of the bright half and the fourth and 10th tithis of the dark half are at the beginning of the first five solar ayanas (solstices). These five are to be repeated for the next five ayanas (solstice points).

Note: A Five-year yuga-cycle has 10 ayanas. So, the above calculation is for the first five ayanas to be

---

[5] Prastha (प्रस्थ) is a Sanskrit unit of weight corresponding to "400 grams" (or, 8 *palas*). It is commonly used in *Rasaśāstra* literature (Medicinal Alchemy) such as the *Rasaprakāśasudhākara* or the *Rasaratna-samuccaya*. Prastha is a weight-unit often used in various Ayurvedic recipes and Alchemical preparations.

repeated in the same order.

Explanation: The solar month consists of 30.5 days and the Synodic month consists of 29.5 days, so there is a difference of 1 tithi. As such a synodic month has 1 more tithi as compared to a solar month. In the ayana there are six synodic months, consequently, it has six tithis more. As such every 7th tithi comes as the beginning of ayana.

If the first ayana starts with the 1st tithi of Śukla Pakṣa; the second solar ayana will start 183 days after the first ayana, i.e. on the 184th day with the 7th tithi Śukla Pakṣa; the third ayana will start 183 days after the second ayana, i.e. on 367th day with 13th tithi of Śukla Pakṣa; fourth will start 183 days after the third, i.e. on 551st day with 4th tithi of Kṛṣṇa Pakṣa; and finally the fifth ayana will start 183 days after the fourth, i.e. on 734th day with 10th tithi of Kṛṣṇa Pakṣa. The same cycle will repeat for another set of 5 ayanas (solstices).

For example:

916th day will start from 1st tithi of Śukla Pakṣa.

1099th day will start from 7th tithi Śukla Pakṣa.

1282nd day will start from 13th tithi of Śukla Pakṣa.

1466th day will start from 4th tithi of Kṛṣṇa Pakṣa.

1649th day will start from 10th tithi of Kṛṣṇa Pakṣa.

This can be verified from the Vedāṅga Calendar revivsed and published by the present author.

Ayanas of the five-year yuga cycle will commence with the following tithis.

Sr. No.	Name of the Year	1st Solstice (Winter Solstice)	2nd Solstice (Summer)
1	Saṁvatsara	Māgha S-1	Śrāvaṇa S-7
2	Parivatsara	Māgha S-13	Śrāvaṇa K-4
3	Idāvatsara	Māgha K-10	Śrāvaṇa S-1
4	Idvatsara	Māgha S-7	Śrāvaṇa S-13
5	Vatsara	Māgha K-4	Śrāvaṇa K-10

Nakṣatras at the beginning of the Ayanas

वसुस्त्वष्टा भवोऽजश्च मित्रः सोऽश्विनौ जलम् ।

धाता कश्चाऽयनाद्याः स्युरर्धपञ्चमभस्त्वृतुः ॥10॥

vasustvaṣṭā bhavo'jaścha mitraḥ so'śvinau jalam;
dhātā kaśchā'yanādyāḥ syurardhapañchamabhastvṛtuḥ.

The 10 ayanas (solstices) of five-year yuga cycle begin with one of the following nakṣatras:

1. First ayana (1st day) with Vasu (Dhaniṣṭhā nakṣtra).
2. Second (184th day) with Tvaṣṭā (Chitrā nakṣtra).
3. Third (367th day) with Bhava (Ārdrā nakṣtra).
4. Fourth (550th day) with Aja (Pūrvabhādrapada nakṣtra).
5. Fifth (733rd day) with Mitra (Anurādhā nakṣtra).
6. Sixth (916th day) with Sarpa (Āśleṣā nakṣtra).
7. Seventh (1099th day) with Aśvinau (Aśvinī nakṣtra).
8. Eighth (1282nd day) with Jala (Pūrvāṣāḍhā nakṣtra).
9. Ninth (1465th day) with Dhātā (Uttaraphālgunī nakṣtra).
10. Tenth (1648th day) with Ka (Rohiṇī nakṣtra).

Explanation: The Moon moves through 1809 nakṣatras in the 10 ayanas of the yuga. So, it moves through about 1809÷10=181 nakṣatras in one ayana. This amounts to 6 sidereal revolutions and 19 nakṣatras. This would make total of 1810 nakṣatras in a yuga, so we have to reduce one nakṣatra at some stage. Lagadha does so at the end of the first ayana. At the end of the first ayana or 12th parva. At the end of the first ayana (12th parva) he drops Puṣya after Punarvasū and uses Āśleṣā, thus making 1809 nakṣatras.

Tithis of the Ṛtus in a Yuga

एकान्तरेऽह्नि मासे च पूर्वान् कृत्वाऽऽदिरुत्तरः ।
अर्धयोः पञ्चवर्षाणामृतु पञ्चदशाऽष्टमे ॥11॥

ekāntare'hni māse cha pūrvān kṛtvā''diruttaraḥ;
ardhayoḥ pañchavarṣāṇāmṛtu pañchadaśā'ṣṭame.

In each of the two halves of the five-year-yuga, the next ṛtu occurs in the alternating periods of one synodic month and tithi after the previous ṛtu. In other words, the consecutive ṛtus occur at intervals of two synodic months and two tithis, because 30 ṛtus (of the five-year-yuga) make up 62 synodic months). The eighth ṛtu (Vasanta) of the 2nd year (Parivatsara) of a five-year-yuga falls on the 15th tithi, i.e. on Pūrṇimā.

Hereunder we give the names of ṛtus and their starting tithis.

1. Śiśira -Magha Śukla 1
2. Vasanta-Chaitra Śukla 3
3. Griṣma-Jyeṣṭha Śukla-5
4. Varṣā-Śrāvaṇa Śukla-7
5. Śarada-Āśvin Śukla-9
6. Hemanta- Mārgaśīrṣa Śukla-11
7. Śiśira -Magha Śukla- 13
8. Vasanta-Chaitra Śukla-15 and so on.

Dropping a day from the fortnight (Tithi Kṣaya)

द्यु हेयं पर्व चेत् पादे पादस्त्रिंशत् तु सैकिका ।

भागात्मनाऽपमृज्यांशान् निर्दिशेदधिको यदि ॥12 ॥

dyu heyaṁ parva chet pāde pādastriṁśat tu saikikā;
bhāgātmanā'pamṛjyāṁ'śān nirdiśedadhiko yadi.

If the end of parva (Amāvasyā or Pūrṇimā) occurs within the first quarter (124÷4=31 tithi aṁśas) or less than one-quarter of the day or say before midday, that tithi is to be omitted from the counting. (That parva will constitute of 14 tihis instead of 15. In other words in such cases 15th tithi (Pūrṇimā or Amāvasyā) will be dropped.

If the tithi aṁśas are more than 31 (that is if the tithi lasts more than one-quarter of the day), the parva whose tithi is to be omitted is found by subtracting its tithi aṁśas from the total no. of tithi aṁśas (i.e. 124) and the if the difference is under 31, the 15th tithi of the given parva will be dropped making the concerned month end on 29th day. In such cases, Darśa (New Moon) and Pūrṇmāsa (Full Moon) iṣṭis will be performed on the following pratipadā.

The above verse has two parts.

1. The first part says that the tithi of a given parva will be dropped if its aṁśas are below 31 (i.e. it lasts less than a quarter of the day). For example, parva 4th ends with 6 tithi aṁśas, since the number is within the first quarter, i.e. 31 tithi aṁśas, so the 15th tithi (Amāvasyā) has been dropped. Similar are the cases of 8th (ending in 10 tithi aṁśas), 12th (ending in 14 tithi aṁśas), 28th (ending in 30 tithi aṁśas) parvas, where 15th tithi (Amāvasyā) will be

dropped owing to the above-cited rule. Conclusively, it can be said that the Amāvasyās of the 14 even parvas 4, 8, 12, 16, 20, 24, 28, 66, 70, 74, 78, 82, 86 and 90 will be dropped. Similarly, Pūrṇimās of 16 odd parvas 33, 37, 41, 45, 49, 53, 57, 61, 95, 99, 103, 107, 111, 115, 119 and 123 will be dropped. It can be confirmed from our Vedāṅga calendar.

2. Second part says that if tithi aṁśas of a given parva are more than 31, subtract those aṁśas from the total tithi aṁśas, i.e. 124. If the difference is below 31, then the 14th tithi will be dropped. For example, the parva 46 ends in 108 tithi aṁśas. If we subtract 108 from 124, we get 124-108=16 which is below 31. So the 14th tithi of 46th parva will be dropped. Similar are the cases of 50th (ending in 112 tithi aṁśas), 54th (ending in 116 tithi aṁśas), 58th (ending in 120 tithi aṁśas), 96th (ending in 96 tithi aṁśas), 108th (ending in 108 tithi aṁśas) and other parvas where 14th tithi will be dropped owing to the above-cited rule.

Explanation: There are 124 tithi aṁśas. The quarter part of 124 is equal to 31, half is 62 and the third part is equal to 93. The Four pādas make it a full day.

Note 1: To find out the tithi aṁśa (parts) at the end of a parva (fortnight), multiply the given parva by 1830 and divide the product by 124. The remainder will be the tithi or day aṁśa at the end of the given parva. For example, if we want to find out the tithi/day aṁśa of parva 14, multiply it by 1830 (14x1830=25620), divide the product, 25620 by 124, we get 206 as quotient and 76 as remainder. The 76 will be the tithi aṁśa at the end of the 14 parva.

Calculation of Parvarāśī

निरेकं द्वादशाभ्यस्तं द्विगुणं चाऽऽप्तसंयुतम् ।
षष्ट्याषष्ट्या युतं द्वाभ्यां पर्वणां राशिरुच्यते ॥13॥

nirekaṁ dvādaśābhyastaṁ dviguṇaṁ chā"ptasaṁyutam;
ṣaṣṭyāṣaṣṭayā yutaṁ dvābhyāṁ parvaṇāṁ rāśiruchyate.

Take the given number of the year in the yuga, subtract this by 1, multiply by 12, again multiply by 2, and add the expired parvas of the given year; if the parva number is more than 60, add 2 for every 60 parvas, and the number obtained is the parva-rāśī (i.e. the total number of parvas expired at the time of calculation).

Mathematically this may be represented as under:

P= (y-1) x 12 x 2 + n + 2 (per 60 parvas) where y is the number of the current year in the yuga and n is the number of parvas elapsed during the year.

Note: In each year there are 12 synodic months, each month has 2 parvas. After 30 months, an intercalary month is added to complete the half yuga. That is why two parvas are added for each expired 60 parvas. Thus we get the total, there are 62 synodic months of 124 parvas in the yuga.

For example, to find the parva-rāśī (total parvas) in the 4th year (Anuvatsara) at end of the Kārtika, Kṛṣṇa aṣṭamī, Anuvatsara being the 4th year of the yuga, so (4-1), we get 3 years, after multiplying 3 by 12 (to convert years into months) we get, 36 and again multiplying 36 by 2 (to convert months into parvas/pakṣas), we get 72 parvas. The present year is Kārtika which is the 10th year.

So years elapsed since Kārtika so far are (10-1) 9. The parvas of 9 expired years will be 9x2=18. Total of no. of parvas are 72+18= 90. The n. of parvas are more than 60, so by adding 2 to 90, we get 92 expired parvas for the 10th Kārtika month of the Anuvatsara (or 4th year of 5-year yuga).

Solar and Lunar Parva-bhaṁśas

Nakṣatra-parts traversed by the Sun and the Moon at the end of a particular Parva

स्युः पादोर्धं त्रिपाद्या यास्त्रिर्द्वयेकेह्नः कृते स्थितिम् ।
साम्येनेन्दोः स्तृणोऽन्ये तु पर्वकाः पञ्चसम्मिताः ॥14॥

syuḥ pādordhaṁ tripādyā yāstrirdvayekehnaḥ kṛte sthitim;
sāmyenendoḥ stṛṇo'nye tu parvakāḥ pañchasammitāḥ.

Since the civil days (1830) of the yuga are divided into quarters (1830/4=457.5, i.e. 457 days and 62 parts), halves (1830/2=915 days) and three quarters (1830/4= 457.5x3=1372 days and 62 parts) corresponding to the divisibility of the Moon's Nakṣatras in the yuga (1809), the Moon's Nakṣatra parts also are taken as a quarter (1809÷4= 452.25, i.e. 452 and 31 parts). Half (1809÷2= 904.5, i.e. 904 and 62 parts) and three quarters (1809÷3=1356.75, i.e. 1356 and 93 parts) without remainder. But the other parts of the Nakṣatras have to be measured in the units of fifth divisions of the parts.

After crossing the parts of the 4th quarter of the 15th tithi, when the following pratipadā (first tithi) maintains its position in the first half of the two halves of the day, any quarter of the three-quarters of pratipadā is qualified for being a parva. There are others who consider pañchakas of nakṣtras corresponding to the position of the Moon qualifying for the status of a parva.

The number of civil days in the yuga, containing 124 parvas, is 1830 (366x5). The parts of the yuga (1830 days) are divided into quarters (1830/4=457.5, i.e. 457 days and 62 parts), halves (1830/2=915 days) and three quarters (1830/4=457.5x3 = 1372 days and 62 parts) corresponding to the Moon's asterismal quarter (1809/4=452 and 31 parts), halves (1809/2=904 and 62 parts) and three quarters (1809/4=452.25x3= 1356 and 93 parts).

Note: The Moon revolves around the Earth 67 times in a five-year-yuga. So, the number of constellations crossed by it are 67x27=1809 in a yuga.

So, at a quarter (31 parvas) Moon's Nakṣatras gone are 1809÷4= 452.25, i.e. 452 and 31 parts; at half (62) of the parvas, the Nakṣatras gone are 1809÷2= 904.5, i.e. 904 and 62 parts. At three quarters (93) of the parvas, the constellations gone are 1809÷3=1356.75, i.e. 1356 and 93 parts.

But at all other parvas, the day-parts being naturally full (since the division of the day into 124 parts is expressed for this purpose), the Nakṣatra-parts cannot be full. So another division of the day called kalās, into 603 parts (1809÷3), is required to express the times of the beginnings of the Nakṣatras. In this unit, the Nakṣatra takes exactly 610 (1830÷3) kalās to pass. It is five times the number of the parts taken by the tithi to pass and we have 122X5=610.

Solar and Lunar Bhāṁśas at the End of a Parva

(Solar & Lunar Bhāṁśas traversed by the Sun and the Moon at the end of a particular Parva)

भांऽशाः स्युरष्टकाः कार्याः पक्षद्वादशकोद्गताः ।

एकादशगुणश्चोनः शुक्लेऽर्धं चैन्दवा यदि ॥15 ॥

bhāṁ'śāḥ syuraṣṭakāḥ kāryāḥ pakṣadvādaśakodgatāḥ;
ēkādaśaguṇaśchonaḥ śukle'rdhaṁ chaindavā yadi.

In a group of 12 pakṣas/parvas, 8 bhāṁśa arise. For the remainder (less than 12 parvas/pakṣas) 11 bhāṁśas (nakṣatra parts) per parva/pakṣa (fortnight) arise. When you calculate lunar bhāṁśa for a parvan ending in full moon-day, add bhāṁśas equal to a half Nakṣatra (62 bhāṁśas)

Mathematically we can say that to find out the Solar and Lunar bhaṁśas (Nakṣatra-parts traversed by the Sun or the Moon) at the end of a particular parva, divide the given parva by 12. Multiply the quotient by 8 and the remainder by 11 and add both. If it is Śukla pakṣa (bright half) and we want to know the bhāṁśas (Nakṣatra-parts) traversed by the Moon, then add half of the bhāṁśa (total of 124 parvas), i.e. 62.

Explanation: the Sun covers 135 (124+11) bhāṁśas, i.e. 1 nakṣatra + 11 bhāṁśas in one parva, so in 12 parvas it covers-

12 x (1 nakṣatra + 11 bhāṁśas)

or   12 nakṣatras + 132 bhāṁśas

or 13 nakṣatras + 8 bhāṁśas

Example: Suppose we want to find out the bhāṁśa

(Nakṣatra-parts) traversed by the Moon and the Sun at the end of 73rd parva, as per instructions

Divide 73 by 12= 6 quotient, 1 remainder

Multiply 6 by 8= 48 and multiply 1 by 11= 11

Add both-48+11=59.

So, we can say that at the end of the 73rd parva, the Sun will traverse 59 bhāṁśas (Nakṣatra-parts).

Since the parva is a bright fortnight, bhāṁśas (Nakṣatra-parts) to be traversed by the Moon will be 59+62=121.

Note:

1. If the number of Bhāṁśa (Nakṣatra-parts) crosses 124, then subtract 124 and the remaining parts will be considered as traversed by the Sun or the Moon as the case may be.

2. If we want to find out the name of the Nakṣatra of the Sun at the end of the 73rd parva, divide the Bhāṁśa (Nakṣatra-part) of that parva by 27 and the remaining will be Nakṣatra of Jāvādi series. For example, divide 59 solar Bhāṁśas of 73rd Parva (Nakṣatra-parts traversed by the Sun at the end of 73rd parva) by 27, the remainder is 5 which shows that the 73rd parva is ending with 59 Nakṣatra-aṁśas (parts) of the 5th, Uttarāṣāḍhā Nakṣatra of the Jāvādi series.

Similarly, if we want to find out the Nakṣatra of the Moon at the end of the 73rd parva, divide lunar bhāṁśas, i.e. 121 by 27 and the remainder 13 shows that the 73rd parva ends with 121 Nakṣatra-parts of the 13th Nakṣatra,

Punarvasu of Jāvādi series.

Note: At the end of even parvas, lunar and solar nakṣatras will be the same, as we have amāvasyas at the end of even parvas and on amāvasyā the Sun and the Moon are at one place or in the same nakṣatra.

The proof of the rule is as follows:

In the yuga of 5 years containing 124 parvas, the Sun traverses 5x27= 135 nakṣatras. Thus during each parva it traverses 135÷124= 1+11/124 (or 1.0887096774) nakṣatras. 11/124 means 11th part of 124. At the end of 12 parvas, it is 12 (1+11/124)= 12+(12x11)/124=12+ 132/124= 12+1+8/124= 13+8/124. 13 in 12 parvas means, it has completed 13 Nakṣatras and 8/124 means, for every 12 parvas it accumulates by 8. For the each remainder, there are 11 bhāṁśas (nakṣatra-parts), so the remainder is multiplied by 11 and added. At the new Moon, the Moon is with the Sun and bhāṁśas (parts) are the same. However, at full moon, the Moon is opposite to the Sun, that is, 13 Nakṣatras and 62 aṁśas (parts) away. So, we add 62 aṁśas (parts) for the Moon.

Below are given Solar and Lunar Parva-bhāṁśas (Nakṣatra-parts traversed by the Sun and the Moon at the end of the Parva).

Parva No.	Solar Nakṣatras and its Añśa (Bhañśa) at the end of the Parva	Lunar Nakṣatras and its Āñśa (Bhāñśa) at the end of the Parva
Saṁvatsara 1	Śatabhiṣak 11°	73 Maghā
2	22 P. Bhādrapada	22 P. Bhādrapada
3	33 U. Bhādrapada	95 Uttaraphalgunī
4	44 Revati	44 Revati
5	55 Aśvinī	117 Chitrā
6	66 Bharaṇī	66 Bharaṇī
7	77 Kṛttikā	15 Anurādhā
8	88 Rohiṇī	88 Rohiṇī
9	99 Mṛgaśirā	37 Mūla
10	110 Ārdrā	110 Ārdrā
11	121 Punarvasu	59 Uttarāṣāḍhā
12	8 Āśleṣā	8 Āśleṣā
13	19 Maghā	81 Dhaniṣṭhā
14	30 P. Phalgunī	30 P. Phalgunī
15	41 U. Phalgunī	103 Pūrvabhādrapada
16	52 Hasta	52 Hasta
17	63 Chitrā	1 Aśvinī
18	74 Svāti	74 Svāti
19	85 Viśākhā	23 Kṛttikā
20	96 Anurādhā	96 Anurādhā
21	107 Jyeṣṭhā	45 Mṛgaśirā
22	118 Mūla	118 Mūla
23	5 Uttrāṣāḍhā	67 Punarvasu
24	16 Śravaṇa	16 Śravaṇa
Parivatsara 25	27 Dhaniṣṭhā	89 Āśleṣā

26	38 Śatabhiṣak	38 Śatabhiṣak
27	49 P. Bhādrapada	111 P. Phalgunī
28	60 U. Bhādrapada	60 U. Bhādrapada
29	71 Revati	9 Chitrā
30	82 Aśvinī	82 Aśvinī
31	93 Bharaṇī	31 Viśākhā
32	104 Kṛttikā	104 Kṛttikā
33	115 Rohiṇī	53 Jyeṣṭhā
34	2 Ārdrā	126-124=2 Ārdrā
35	13 Punarvasu	75 Pūrvāṣāḍhā
36	24 Puṣya	24 Puṣya
37	35 Āśleṣā	97 Śravaṇa
38	46 Maghā	46 Maghā
39	57 P. Phalgunī	119 Śatabhiṣak
40	68 U. Phalgunī	68 U. Phalgunī
41	79 Hasta	17 Revatī
42	90 Chitrā	90 Chitrā
43	101 Svāti	39 Bharaṇī
44	112 Viśākhā	112 Viśākhā
45	123 Anurādhā	61 Rohiṇī
46	10 Mūla	10 Mūla
47	21 Pūrvāṣāḍhā	83 Ārdrā
48	32 Uttarāṣāḍhā	32 Uttarāṣāḍhā
Idāvatsara 49	43 Śravaṇa	105 Puṣya
50	54 Dhaniṣṭhā	54 Dhaniṣṭhā
51	65 Śatabhiṣak	127-124=3 P. Phalgunī
52	76 P. Bhādrapada	76 P. Bhādrapada

53	87 U. Bhadrapada	25 Hasta
54	98 Revati	98 Revati
55	109 Aśvanī	47 Svāti
56	120 Bharaṇī	120 Bharaṇī
57	7 Rohiṇi	69 Anurādhā
58	18 Mṛgaśirā	18 Mṛgaśirā
59	29 Ārdrā	91 Mūla
60	40 Punarvasū	40 Punarvasū
61	51 Puṣya	113 Uttarāṣāḍhā
62	62 Āśleṣā	62 Āśleṣā
63	73 Maghā	11 Śatabhiṣak
64	84 P. Phalgunī	84 P. Phalgunī
65	100 U.Phalgunī	33 U. Bhādrapada
66	106 Hasta	106 Hasta
67	122 Chitrā	55 Aśvinī
68	4 Viśākhā	128-124=4 Viśākhā
69	15 Anurādhā	77 Kṛttikā
70	26 Jyeṣṭhā	26 Jyeṣṭhā
71	42 Mūla	99 Mṛgaśirā
72	48 Pūrvāṣāḍhā	48 Pūrvāṣāḍhā
73	59 Uttarāṣāḍhā	121 Punarvasu
74	70 Śravaṇa	70 Śravaṇa
Anuvatsara 75	81 Dhaniṣṭhā	19 Maghā
76	92 Śatabhiṣak	92 Śatabhiṣak
77	103 P. Bhādrapada	41 U. Phalgunī
78	114 U. Bhādrapada	114 U. Bhādrapada
79	1 Aśvinī	63 Chitrā

80	12 Bharaṇī	12 Bharaṇī
81	23 Kṛttikā	85 Viśākhā
82	34 Rohiṇī	34 Rohiṇī
83	45 Mṛgaśira	107 Jyeṣṭhā
84	56 Ārdrā	56 Ārdrā
85	67 Punarvasū	5 Uttarāṣāḍhā
86	78 Puṣya	78 Puṣya
87	89 Āśleṣā	27 Dhaniṣṭhā
88	100 Maghā	100 Maghā
89	111 P. Phalgunī	49 P. Bhādrapada
90	122 U. Phalgunī	122 U. Phalgunī
91	9 Chitra	71 Revatī
92	20 Svāti	20 Svāti
93	31 Viśākhā	93 Bharaṇī
94	42 Anurādhā	42 Anurādhā
95	53 Jyeṣṭhā	115 Rohiṇī
96	64 Mūla	64 Mūla
97	75 Pūrvāṣāḍha	13 Punarvasu
98	86 Uttarāṣāḍhā	86 Uttarāṣāḍhā
Idavastsara 99	121 Śravaṇa	35 Āśleṣā
100	108 Dhaniṣṭhā	108 Dhaniṣṭhā
101	19 Śatabhiṣak	57 P. Phalgunī
102	6 U. Bhādrapad	6 U. Bhādrapad
103	41 Revati	79 Hasta
104	28 Aśvinī	28 Aśvinī
105	63 Bharaṇī	101 Svāti
106	50 Kṛttikā	50 Kṛttikā

107	85 Rohiṇī	123 Anurādhā
108	18 Mṛgaśira	18 Mṛgaśira
109	83 Ārdrā	21 Pūrvāṣāḍha
110	94 Punarvasu	94 Punarvasu
111	105 Puṣya	43 Śravaṇa
112	116 Āśleṣā	116 Āśleṣā
113	3 P. Phalgunī	65 Śatabhiṣak
114	14 U. Phalgunī	14 U. Phalgunī
115	25 Hasta	87 U. Bhādrapada
116	36 Chitrā	36 Chitrā
117	47 Svāti	109 Aśvinī
118	58 Viśākhā	58 Viśākhā
119	69 Anurādhā	7 Rohiṇī
120	80 Jyeṣṭhā	80 Jyeṣṭhā
121	91 Mūla	29 Ārdrā
122	102 Pūrvāṣāḍhā	102 Pūrvāṣāḍhā
123	113 Uttarāṣāḍhā	51 Puṣya
124	124 Śravaṇa	124 Śravaṇa

Hour Angle and Lagna (Ascendant) of a Nakṣatra at the end of a Particular Parva

नवकैरुद्गतोंऽशः स्यादूनः सप्तगुणो भवेत् ।
आवापस्त्वयुजेऽर्धं स्यात् पौलस्त्येऽस्तङ्गतेऽपरम् ॥16॥

navakairudgatoṁ'śaḥ syādūnaḥ saptaguṇo bhavet;
āvāpastvayuje'rdhaṁ syāt paulastye'staṅgate'param.

To know the hour angle and lagna of a Nakṣatra at the end of parva, divide the number of parvas by 9. Each of the remainders should be multiplied by 7 and add to the quotient. If the quotient is odd, add half (that is 62 parts). If the Paulastya (Moon) is setting when the sun rises (i.e. of full moon parva), add another half (i.e. 62 parts).

For example, if we want to know the hour angle and lagna (ascendant) of a Nakṣatra at the end of the 93rd parva, divide the 93rd parva by 9, and we get 10 as quotient and 3 as remainder. Multiply the remainder by 7, i.e. 3x7=21 and add to the quotient (10 + 21 = 31). 93 is full moon parva, so add 62 to 31 (62+31= 93). As such we are able to know that by the end of the 93rd parva, the 93rd bhāṁśas (parts of nakṣatra) is the rising point.

If we want to know the name of nakṣatra, subtract from bhṁśa of 93rd parva the multiples of 27, i.e. 93-(27X3)=12. It shows that at the end of the 93rd parva, 93 aṁśas of the 12th nakṣatra named Bharaṇī of the Jāvādi list is the rising point, lagna.

Note: Aṁśa (part) of tithi on a particular day will be known as the hour angle of the Sun. By subtracting the hour angle of the Sun (tithi aṁśas) from the nakṣtra-aṁśas obtained, we get the hour angle of Dhaniṣṭhā.

For example, by subtracting the Sun's hour angle from obtained bhāṁśas (nakṣatra aṁśas), we get 93-62=31 as the hour angle of Dhaniṣṭhā. 31x27÷124=6.75, i.e. 6th nakṣatra counted from Dhaniṣṭha, i.e. Bharaṇi is again found as rising, lagna at the end of the 93rd parva.

Jāvādi List of Nakṣatras

जावाद्यंशैः समं विद्यात् पूर्वार्धे पर्वसूत्तराः ।
भादानं स्याच्चतुर्दश्यां द्विभागेभ्योऽधिको यदि ॥17॥

jāvādyaṁśaiḥ samaṁ vidyāt pūrvārdhe parvasuttarāḥ;
bhādānaṁ syāchchaturdaśyāṁ dvibhāgebhyo'dhiko yadi.

Know that the Jāvādi list is balanced. The former letter in the list represents the lunar nakṣatra at the beginning of the first half. If the parva falls within the first half of the bhāṁśa, i.e. if the bhāṁśa is 62 or less at parva, the beginning of the Nakṣatra (bhādānam) will fall in the parva tithi, i.e. 15th tithi itself. If the Nakṣatra part is greater than the parts of the day at which the parva falls, the beginning of the Nakṣatra falls on the Chaturdaśī tithi day.

Example:

१जौ २द्रा ३गः ४खे ५श्वे ६ही ७रो ८षाश् ९चिन् १०मू ११षक १२ण्यः १३सू १४मा १५धा १६णः ।

१७रे १८मृ (म्रे) १९घाः २०स्वा २१पो २२जः २३कृ २४ष्यो २५ह २६ज्ये २७ष्टा इत्यृक्षा लिङ्गैः ॥18॥

1 jau, 2 drā, 3 gaḥ, 4 khe, 5 śve, 6 hī, 7 ro, 8 ṣāś, 9 chin, 10 mū, 11 ṣaka, 12 ṇyaḥ, 13 sū, 14 mā, 15 dhā, 16 ṇaḥ;

17 re, 18 mṛ (mre), 19 ghāḥ, 20 svā, 21 po, 22 jaḥ, 23 kṛ, 24 ṣyo, 25 ha, 26 jye, 27 ṣṭhā, ityṛkṣā liṅgaiḥ.

Take the Nakṣatras represented symbolically in Jāvādi list in order of 1,2,3 (which is different from the normal order of the Nakṣatras). So many aṁśas (parts) of that Nakṣatra have gone at the end of that parva for which that bhāṁśa has been found.

Explanation: In a year, the Moon goes ahead of the Sun by 27+10 nakṣatras or say 13.5 + 5 nakṣatras in an ayana (semester). The five nakṣatras difference in Jāvādi list is attributed to this circumstance.

Nakṣatra Table with Symbols

No	Symbol	Nakṣatra	No	Symbol	Nakṣatra
1	Jau	Āśvayujau	14	Mā	Aryamā (Uttaraphalgunī)
2	Dra	Ārdrā	15	Dhāḥ	Anurādhāḥ
3	Gaḥ	Bhagḥ (Pūrva Phalgunī)	16	Ṇaḥ	Śravaṇaḥ
4	Khe	Viśākhe	17	Re	Revati
5	Śve	Viśvedevāḥ (Uttarāṣāḍhā)	18	Mṛ	Mṛgaśirāḥ
6	Hiḥ	Ahirbudhnyaḥ (U. Bhādrapada)	19	Ghaḥ	Maghaḥ
7	Ro	Rohiṇī	20	Svā	Svāti
8	Ṣā	Āśleṣā	21	Paḥ	Āpaḥ (Pūrvāṣāḍhā)
9	Cit	Chitrā	22	Jaḥ	Ajaekapāt (P. Bhādrapada)
10	Mū	Mūla	23	Kṛ	Kṛttikāḥ
11	Śa	Śatabhiṣak	24	Ṣyaḥ	Puṣyaḥ
12	Ṇyaḥ	Bharaṇyaḥ	25	Ha	Hastaḥ
13	Su	Punarvasu	26	Jye	Jyeṣṭhā
			27	Ṣṭhāḥ	Śraviṣṭhāḥ (Dhaniṣṭhā)

Note: In the Javādi list, Nakṣatras have been arranged from Aśvinī, each being the sixth from the normal arrangement of Nakṣatras.

The normal order of the Nakṣatras reckoned from Aśvinī is given below:

1. Aśvinī
2. Bharaṇi
3. Kṛttikā
4. Rohiṇī

5.	Mṛgaśirā	6.	Ārdrā
7.	Punarvasu	8.	Puṣya
9.	Aśleṣā	10.	Maghā
11.	Pūrva Phalguni	12.	Uttara Phalguni
13.	Hasta	14.	Chitrā
15.	Svāti	16.	Viśākhā
17.	Anurādhā	18.	Jyeṣṭhā
19.	Mūla	20.	Pūrvāṣāḍhā
21.	Uttarāṣāḍhā	22.	Śrāvaṇa
23.	Dhaniṣṭhā	24.	Śatabhiṣak
25.	Pūrva Bhādrapada	26.	Uttara Bhādrapada
27.	Revati		

Order of the Nakṣatras reckoned from Dhaniṣṭhā (Śraviṣṭhā) is given below:

1.	Dhaniṣṭhā	2.	Śatabhiṣak
3.	Pūrva Bhādrapada	4.	Uttara Bhādrapada
5.	Revati		
6.	Aśvinī	7.	Bharaṇi
8.	Kṛttikā	9.	Rohiṇi
10.	Mṛgaśirā	11.	Ārdrā
12.	Punarvasu	13.	Puṣya
14.	Aśleṣā	15.	Maghā
16.	Pūrva Phalgunī	17.	Uttara Phalgunī
18.	Hasta	19.	Chitrā
20.	Svāti	21.	Viśākhā
22.	Anurādhā	23.	Jyeṣṭhā
24.	Mūla	25.	Pūrvāṣāḍhā
26.	Uttarāṣāḍhā	27.	Śrāvaṇa

Lunar/Solar Parva Bhādānakalā

(Solar/Lunar Nakṣatra Start Time in Kalās at the end of a Parva)

कार्या भांशाष्टकस्थाने कला एकान्नविंशतिः ।

ऊनस्थाने तु (त्रि) सप्ततिमुद्वपेद् युक्तिसम्भवे ॥19 ॥

kāryā bhāṁśāṣṭakasthāne kalā ēkānnaviṁśatiḥ;
ūnasthāne tu (tri) saptatimudvaped yuktisambhave.

To know the bhādānakalā of the Moon or the Sun (Lunar or Solar Nakṣatra start time in kalās) at the end of a given parva, substitute 19 kalās for every group of 8 bhaṁśas and take out from them 73 times the remainder bhaṁśas (when the remainder occurs).

Mathematically saying, divide the bhāṁśas at the end of parva by 8, multiply the quotient by 19 and take out the remainder multiplied by 73 and you will get Bhādānakalā of Moon at the end of a parva.

$B^0/8 = q19 - r73$

For example, if we want to find out the bhādānakalā of the Moon at the end of the 6th parva. We have 66 bhāṁśas in the Moon's position at the end of the 6th parva. Divide 66÷8, we have quotient 8 and remainder 2 bhaṁśas. After applying the above rule, we have (8 x 19 - 2 x 73) = 6 bhadānakalās at the end of the 6th parva. That is the time of Moon's entry into the New Moon position at the end of the 6th parva.

Explanation: The Quotient is multiplied by 19 because 8 bhāṁśas arise in 12 parvas during which the moon moves 175 nakṣatras and 8 bhāṁśas. Now the Bhādānakalās of 175 nakṣatras are 175 x 7=1225.

Subtracting 1206 kalās of 2 full days (603 x 2), we get a remainder of 19. Multiplication of remainder by 73 is because the first parva has 73 parva-bhāṁśas.

Nakṣatra of a Tithi as per the Jāvādi List

तिथिमेकादशाभ्यस्तां पर्वभांशसमन्विताम् ।
विभज्य भसमूहेन तिथिनक्षत्रमादिशेत् ॥20 ॥

tithimekādaśābhyastāṁ parvabhāṁśasamanvitām;
vibhajya bhasamūhena tithinakṣatramādiśet.

Multiply the required tithi of the parva by 11, and adding it to the parts of the parva (parva-bhāṁśa) of the expired parva, and divide out by the total number of Nakṣatras (27) if the sum is greater than 27. The remainder will be the Nakṣatra listed in the Javādi series of the required tithi.

For instance, to find a Nakṣatra of the 6th tithi after the 7th parva:

Multiply 6 by 11 = 66, and add to it the Nakṣatra part of expired parva 7, which is 15. The Total is 66+15=81. 81 is greater than 27, so divide it by 27, the remainder is 0. Applying the Jāvādi series, the 27th, Dhaniṣṭhā is the Nakṣatra of the 6th tithi of the 8th parva as per the Vedāṅga Jyotiṣa calendar.

Note: In the Vedāṅga Jyotiṣa calendar, the Nakṣatra next to a given Nakṣatra occurs 11 places away, so tithis are multiplied by 11.

Tithi Bhādānikā/Bhādānakalā

(Tithi Nakṣatra End time in Kalās)

याः पर्वभादानकलास् तासु सप्तगुणा तिथिः ।

युक्ता ताश्च विजानीयात् तिथिभादानिकाः कलाः ॥21॥

yāḥ parvabhādānakalās tāsu saptaguṇā tithiḥ;
yuktā tāścha vijānīyāt tithibhādānikāḥ kalāḥ.

Add elapsed tithi multiplied by 7 to the bhādāna kalā (beginning of the Nakṣatra) of the elapsed parva, we get the Bhādānakalā (Nakṣatra start time) of tithis.

Explanation: The duration of the Moon in a Nakṣatra is 610 kalās and the duration of the Moon in a tithi is 603 kalās, so a Nakṣatra begins 7 kalās later daily as compared to the Bhādāna-kalā of a tithi. Hence the rule of adding 7 times the tithis, so that Nakṣatra may come in consonance with a Parvabhādāna -kalā.

Example: To find the Bhādāna-kalā of the fifth tithi of 13th parva, multiplying elapsed tithi by seven, we get-

$4 \times 7 = 28$

And adding it to the bhādānakalā of the elapsed parva 12, i.e. 19, we get-

$28 + 19 = 47$

Similarly, if we want to find out the 8th tithi of the 93rd parva, multiplying 7 elapsed tithis by 7, we get

$7 \times 7 = 49.$

Add Bhādānakalā of the elapsed parva 92, 356 to it and we get the Bhādāna-kalā of 8th tithi of 93rd Nakṣatra as $49 + 356 = 405.$

End Time of the Tithi in Aṁśas and Diurnal Position of the Sun

अतीतपर्वभागेभ्यः शोधयेद् द्विगुणां तिथिम्।

तेषु मण्डलभागेषु तिथिनिष्ठां गतो रविः ॥22॥

atītaparvabhāgebhyaḥ śodhayed dviguṇāṁ tithim;
teṣu maṇḍalabhāgeṣu tithiniṣṭhāṁ gato raviḥ.

Multiply the given tithi by 2 and subtract it from the tithi aṁśa (time of the end of the tithi in aṁśas) ending the previous parva. This is the same as the position of the Sun in the Nāḍimaṇḍal (diurnal circle).

If we want to know the end time of the 8th tithi after the 93rd parva, subtract 8x2=16 from the tithi aṁśa (part of the day in aṁśas), 92 ending the previous (92nd) parva. The result is 92-16=76.

The Day of Vernal Equinox (Viṣuva)

विषुवन्तं द्विरभ्यस्तं रूपोनं षड्गुणीकृतम् ।

पक्षा यदर्ध पक्षाणां तिथिः स विषुवान् स्मृतः ॥23॥

viṣuvantaṁ dvirabhyastaṁ rūponaṁ ṣaḍguṇīkṛtam;
pakṣā yadardha pakṣāṇāṁ tithiḥ sa viṣuvān smṛtaḥ.

Take the number of viṣuva (equinox) and multiply it by 2; subtract 1; and multiply the balance by 6. What is obtained is the number of elapsed parvas of the viṣuv. [This may be represented mathematically as $P=(2v-1)\times6$] Half of this is the tithi at the end of which Viṣuva (equinox) occurs. This may be represented mathematically as:

$$T=(v-1/2)\times6$$

Note: In Vedāṅga Jyotiṣa viṣuva is taken to be the middle point of the ayana. Hence, there are two viṣuvas: one is Vernal Equinox (Vasanta Sampāt) and the second is Autumnal Equinox (Śarad Sampāt). The first viṣuva occurs after 93 tithis or say 6 parvas and 3 tithis.

Thereafter the viṣuvas repeat after 186 tithis or 12 parvas and 6 tithis.

There are 10 equinoxes in a five-year yuga of Vedāṅga Jyotiṣa, every year there are two equinoxes. The vernal equinox will be numbered one and Autumn equinox will be numbered two in the first year, the second year equinoxes will be 3rd and 4th, so on up to 10th.

Example: To know the tithi of the 3rd equinox, multiply 3 by 2=6. Subtract one 6-1=5, and multiply the balance 5 by 6=30. 30 will represent here elapsed parvas. Half of 30, i.e. 15 will be the 3rd Viṣuvān tithi or the day of the 3rd equinox (vernal equinox). Hereunder we furnish a table depicting tithis of various viṣuvas (Vernal and Automnal Equinoxes).

Year	Tithis of Vernal Equinox	Automnal Equinox
1. Saṁvasara	1. Vaiśākha S-3	2. Kārtika S-9
2. Parivatsara	3. Vaiśākha S-15	4. Kārtika K-6
3. Idāvatsara	5. Vaiśākha K-12	6. Kārtika S-3
4. Anuvatsara	7. Vaiśākha S-9	8. Kārtika S-15
5. Idvatsara	9. Vaiśākha K-6	10. Kārtika K-12

Time Measures

पलानि पञ्चाशदपां धृतानि तदाढकं द्रोणमतः प्रमेयम् ।
त्रिभिर् विहीनं कुडवैस्तु कार्यं तन्नाडिकायास्तु भवेत् प्रमाणम् ॥24॥

palāni pañchāsadapāṁ dhṛtāni tadāḍhakaṁ droṇamataḥ prameyam;
prameyam;

tribhir vihīnaṁ kuḍavaistu kāryaṁ tannāḍikāyāstu bhavet pramāṇam.

50 palas of water is the measure called āḍhaka. From this is derived the measure of droṇa (which is four times the āḍhaka). This decreased by three kuḍava measures (i.e. three sixteen 3/16 of an āḍhaka) is the volume measured for the length of one nāḍikā of time.

The above measurement of the *Vedāṅga Jyotiṣa* is further clarified by the *Skanda Purāṇa* (*Kāśī Khaṇḍa*, 27.148) as under:

पलं च कुडवः प्रस्थ आढको द्रोण एव च ।
धान्यमानेन बोद्धव्याः क्रमशोऽमी चतुर्गुणाः ।।

palaṁ cha kuḍavaḥ prastha āḍhako droṇa ēva cha;
dhānyamānena boddhavyāḥ kramaśo'mī chaturguṇāḥ.

That is pala, kuḍava, prastha, āḍhaka and droṇa each measures are 4 times of the following ones.

4 palas	= 1 kuḍava
4 kuḍavas	= 1 Prastha (16 palas)
4 prasthas	= 1 āḍhaka (50 palas)
4 āḍhakas	= 1 droṇa (200 palas)

Sun's Nakṣatras and Bhāṁśas

एकादशभिरभ्यस्य पर्वाणि नवभिस्तिथिम् ।

युगलब्धं सपर्व स्याद् वर्तमानार्कभं क्रमात् ॥25॥

ēkādaśabhirabhyasya parvāṇi navabhistithim;
yugalabdhaṁ saparva syād vartamānārkabhaṁ kramāt.

Multiplying the elapsed parvas by 11, and previous tithis by 9; adding the two and dividing the total by the number of parvas in the yuga; and adding the number of elapsed parvas to the quotient, we get the parts (aṁśas) and Nakṣatra of the Sun.

Example: To find out the Sun's Nakṣatra and parts in the 9th tithi of the 94th parva, multiply the elapsed parvas, 93 by 11 (93x11=1023); and elapsed tithis, 8 by 9 (8x9=72). Adding the two 1023+72, we get 1095. Dividing the 1095 by total no. of parvas in a 5-year yuga, i.e. 124 (1095÷124) we get 8 full Nakṣatras elapsed as quotient and 103 as remainder which shows that the Sun is at 103 aṁśas. Adding the quotient 8 to the elapsed Parvas 93, we get 101. Diving 101 by 27 we find the remainder 20 which shows that the Sun is at the 103rd part of 20th Nakṣatra from Dhaniṣṭhā or 21st Nakṣatra is starting from Dhaniṣṭhā, i.e. Viśākhā.

Day and Aṁśas of Sun's Nakṣatra

सूर्यर्क्षभागान् नवभिर्विभज्य शेष द्विरभ्यस्य दिनोपभुक्तिः ।

तिथेर्युता भुक्तिदिनेषु कालोऽयोगे दिनैकादशकेन तद् भम् ॥26॥

sūryarkṣabhāgān navabhirvibhajya śeṣa dvirabhyasya
dinopabhuktiḥ;
titheryutā bhuktidineṣu kālo'yoge dinaikādaśakena tad
bham.

Divide the aṁśa of the Sun's nakṣatra (got from the rule cited in śloka 25) by 9 and multiply the quotient by 2. This product is the additional aṁśas (parts) traversed by the Sun. Adding the product to the tithi aṁśas of the completed tithi, we get the tithi aṁśas of the entry of the Sun into the nakṣatra. Adding the product to the nakṣatra aṁśas and dividing the sum by 11, we find the days before the completed tithi when the Sun entered the nakṣatra.

Example: To find the day and the aṁśas of the Sun's entry into the given nakṣatra on the 9th tithi of 94th parva.

As known from the example of the śloka no. 25th above, Sun's Nakṣatra in the 9th tithi of the 94th parva is Viśākhā and Nakṣatra aṁśa is 103.

So, as per the above rule, by dividing Sun's nakṣatra-aṁśa 103 by 9, we get the quotient 11.4 aṁśa. Multiplying the quotient 11.4 by 2, we get 22.8. Adding 22.8 aṁśas to the tithi aṁśas, 46, of the completed 8th tithi we get 68.8, which shows that Viśākhā began when 68 tithi aṁśas was completed.

Similarly, by adding 22.8 aṁśas to the nakṣatra-aṁśas, 103, of the 94th parva, we get 125.8 aṁśas and by dividing the same by 11, we get 11.4 which shows that Viśākhā began 11 days before the 8th tithi of 94th parva, i.e. on the 12th tithi of 93rd parva when 68 tithi-aṁśa has completed.

Explanation: Nakṣatra-aṁśas divided by 9 gives the tithi periods for the aṁśas to go. The number of tithis can be taken as the number of days minus twice the number of tithis because each tithi is two aṁśas less than a day. But these days have to be subtracted from the given day and aṁśas. So twice the number of tithis are added to the aṁśas of the day and the number of the tithis are subtracted from the point of the added aṁśas of the day.

Correction for the Sidereal Day & Five Year Yuga Scheme

त्र्यंशो भशेषो दिवसांशभागश्चतुर्दशस्याऽप्युपनीय भिन्नम् ।
भार्धाऽधिके चापि गते परोंऽशो द्वावुक्तमेकं नवकै रवेर्द्यु ॥27 ॥

trayaṁśo bhaśeṣo divasāṁśabhāgaśchaturdaśasyā'pyupanīya bhinnam;
bhārdhā'dhike chāpi gate paroṁ'śo dvāvuktamekaṁ navakai raverdyu.

The nakṣatra in the sidereal circle is i/3rd bhāṁśa more than the diurnal circle. That is, the sidereal circle has 124 bhāṁśas, but the diurnal circle has 124-1/3=124 2/3 or 123.6666666667 bhāṁśas. To bring the diurnal circle in unison with the sidereal circle, add the excess 9 bhāṁśas (or 1 tithi) arising out of the excess of 1/3rd bhāṁśas mentioned above to the 14th tithi as the 15th tithi at the bhārdha (first half) parva (62nd parva) or the prāṁśa (end of the 2nd half) of the yuga, i.e. 124th parva (wherever the 15th tithi is dropped). Thus the above rule talks about the deletion of 29 tithis instead of 30 tithis as ruled in śloka 12.

Explanation: The Sidereal circle has 124 x 27=3348 bhāṁśas in one year and the diurnal circle will have 123.6666666667 x 27= 3339 bhāṁśa in a year. Thus

3348 - 3339 = 9 bhāṁśas will be in excess each year. 9 bhāṁśas are equal to 1 tithi. One tithi has 122 aṁśas. To bring the sidereal circle in unison with the diurnal circle this excess of 9 bhāṁśas or one tithi is to be added either when the sun crosses the first half number of parvas, i.e. at the end of 62nd parva and second-half number of parvas, i.e. 124th parva. Adding one tithi means retaining the 15th tithi if it is being dropped either at the end of 62nd parva or at the end of 124th parva as per rule prescribed in śloka no. 12 of the Yājuṣa Jyotiṣa. We find in the calendar that the 15th tithi is dropped from both the parvas. So we can retain the 15th tithi either of 62nd parva or the 124th parva

Explanation 1: One sidereal circle consisting of 367 days has 3348÷9=372 tithis and a diurnal circle consisting of 366 days has 3339÷9= 371 tithis. As such, there is a difference of 1 tithi. We are therefore required to add 3348-3339=9 bhāṁśas or 367-366=1 tithi.

Note: Here Chaturdaśī is mentioned for any addition or at the end of the parva because the 15th tithis (amāvasyās and pūrṇimās) are often dropped as kṣaya tithis.

Explanation 2: We are informed by the Vedāṅga Jyotiṣa that the 5-year Yuga contains:

1. The 60 solar months of 30.5 days each, which makes a Yuga of 60 x 30.5= 1830 solar days.

2. The 61 sāvana months of 30 days each, which amounted to 61 x 30= 1830 sāvana days in a Yuga, requiring one adhikamāsa in 5 years for being in unison

with solar months.

3. The 62 synodic lunar months (tithimāsas) of 29.532 days each which makes a Yuga of 62 x 29.532=1830.984 or 1831 days, requiring two adhikamāsas in 5 years for being in unison with solar months.

4. The 67 sidereal lunar months (nakṣatra months) of 27.32 days each, which amounted to 67 x 27.32=1830.44 days, requiring 7 adhikamāsās in 5 years years for being in unison with solar months.

In other words, 1830 is the number of days in one Yuga represented a least common multiple (LCM) of solar, sāvana, synodic luna and sidereal lunar months. However in reality a yuga contained 1831 days. So, the length of the Yuga has to be increased by one day or 124 aṁśas in five years. By retaining one tithi at the end of the 62nd or 124th parva we may add 122 aṁśas. The rest of the two aṁśas may also be added at the end of the 124th parva. So the 15th tithi of 124th parva will have 124 aṁśas.

Time Measures

त्रिशत्यहां सषद्दष्टिरब्दः षट् चर्तवोऽयने ।
मासा द्वादश सौराः स्युरेतत् पञ्चगुणं युगम् ॥28॥

triśatyahāṁ saṣaṭṣaṣṭirabdaḥ ṣaṭ chartavo'yane;
māsā dvādaśa saurāḥ syuretat pañchaguṇaṁ yugam.

366 days make one year. In a solar year, there are six ṛtus and two ayanas (Uttarāyaṇa & Dakṣiṇāyana). In a solar year, there are 12 solar months (30.5 days each). The five solar years make a yuga.

Yuga and Its Elements

उदया वासवस्य स्युर्दिनराशिः सपञ्चकः ।

ऋषेर्द्विषष्ट्या हीनः स्याद् विंशत्या सैकया स्तृणाम् ॥29 ॥

udayā vāsavasya syurdinarāśiḥ sapañchakaḥ;
ṛṣerdviṣaṣṭyā hīnaḥ syād viṁśatyā saikayā stṛṇām.

The number of rising of Nakṣatra Dhaniṣṭhā in the yuga is the number of days, 1830 plus five (1830+5=1835). It means that a yuga has 1835 sidereal days. The number of rising of Dhaniṣṭhā Nakṣatra in one year is 1835÷5=367. It means that a sidereal year has 367 days. The number of rising of the Moon is the number of days, 1830 minus 62 (1830-62=1768). The total number of rising of the Moon's 27 Nakṣatras in a yuga is the number of solar days, 1830 minus 21 (1830-21=1809).

The other proof is the Moon revolves around the Earth on her ecliptic path constituted of 27 Nakṣatras 67 times. This also makes 67x27=1809.

पञ्चत्रिंशं शतं पौष्णमेकोनमयनान्यृषेः ।

पर्वणां स्याच्चतुष्पादी काष्ठानां चैव ताः कलाः ॥30 ॥

pañchatriṁśaṁ śataṁ pauṣṇamekonamayanānyṛṣeḥ;
parvaṇāṁ syāchchatuṣpādī kāṣṭhānāṁ chaiva tāḥ kalāḥ.

In the same way, the total number of Nakṣatra traversed by the Sun in a Yuga is 135. (The earth completes five revolutions around the Sun on her ecliptic path constituted of 27 Nakṣatras in five years. So, also this makes 27 x 5= 135). In a 5-year yuga, the number of lunar parvas/pakṣas are equal to the sum of four quarters (31 x 4= 124). The same number of kāṣṭhās, i.e. 124 kāṣṭhas make one kalā.

सावनेन्दुस्तृमासानां षष्टिः सैकद्विसप्तिका ।
द्युस्त्रिंशत् सावनः साऽर्धः सौरः स्तृणां स पर्ययः ॥31॥

sāvanendustṛmāsānāṁ ṣaṣṭiḥ saikadvisaptikā;
dyustrimśat sāvanaḥ sā'rdhaḥ sauraḥ stṛṇāṁ sa paryayaḥ.

There are 60+1,2,7 (i.e. 61, 62 and 67) Sāvana months, Lunar (Synodic) months and Lunar (Sidereal) months respectively in a yuga. The 61 Sāvana months contain 30 days each (or 61x30=1830 days in a 5-year yuga or 366 days in a year); 30.5 days make the Solar month (We can also say that in a 5-year yuga there are 60 solar months of 30.5 days each). The synodic month has 62x30=1860 days in 5-year Yuga or 1860÷5=372 tithis in a year); and one Sidereal month is made by the Moon's one revolution around the Nakṣatras. This may be calculated as 1830÷67=27.31 days.

Note: Lunar synodic month contains 354 days in a year or 29.531 days in a month. This may also be worked out in the language of Vedāṅga Jyotiṣa as 1830÷62=29.516 days.

Vedic Names of the Nakṣatras

अग्निः प्रजापतिः सोमो रुद्रोऽदितिर् बृहस्पतिः ।
सर्पाश्च पितरश्चैव भगश्चैवाऽर्यमाऽपि च ॥32॥

agniḥ prajāpatiḥ somo rudro'ditir bṛhaspatiḥ;
sarpāścha pitaraśchaiva bhagaśchaivā'ryamā'pi cha.

सविता त्वष्टाऽथ वायुश्चेन्द्राग्नी मित्र एव च ।
इन्द्रो निर्ऋतिरापो वै विश्वेदेवास्तथैव च ॥33॥

savitā tvaṣṭā'tha vāyuśchendrāgnī mitra ēva cha;
indro nirṛtirāpo vai viśvedevāstathaiva cha.

विष्णुर्वसवो वरुणोऽज एकपात् तथैव च ।

अहिर्बुध्यस्तथा पूषा अश्विनौ यम एव च ॥34॥

viṣṇurvasavo varuṇo'ja ēkapāt tathaiva cha;
ahirbudhnyastathā pūṣā aśvinau yama ēva cha.

The Devatas (Vedic names) of various Nakṣatras are as
under:

Sr. No.	Devata (Vedic Name)	Name of Nakṣatra
1	Agni	Kṛttikā
2	Prajāpati (Brahmā)	Rohiṇī
3	Soma	Mṛgaśira
4	Rudra	Ārdrā
5	Aditi	Punarvasu
6	Bṛhaspati	Tiṣya
7	Sarpa	Āśleṣā
8	Pitara	Maghā
9	Bhaga	Purva Phālgunī
10	Aryamā (Dhātā)	Uttara Phalgunī
11	Savitā	Hasta
12	Tvaṣṭā	Chitrā
13	Vāyu	Svāti
14	Indrāgnī	Viśākhā
15	Mitra	Anurādhā
16	Indra	Jyeṣṭhā
17	Nirṛti	Mūla
18	Āpaḥ	Pūrvāṣāḍhā
19	Viśvedeva	Uttarāṣāḍhā
20	Viṣṇu	Śravaṇa
21	Vasu	Dhaniṣṭhā
22	Varuṇa	Śatabhiṣak
23	Aja Ekapāt	Pūrvabhādrapada

24	Ahirbudhnya	Uttara Bhādrapada
25	Pūṣā	Revatī
26	Aśvinīkumāra	Aśvinī
27	Yama	Bharaṇī or Apabharaṇī

Naming based upon Vedic Names of the Nakṣatras

नक्षत्रदेवता ह्येता एताभिर्यज्ञकर्मणि ।

यजमानस्य शास्त्रज्ञैर्नाम नक्षत्रजं स्मृतम् ॥35॥

nakṣatradevatā hyetā ētābhiryajñakarmaṇi;
yajamānasya śāstrajñairnāma nakṣatrajaṁ smṛtam.

These are the Devatās (Vedic names) of the Nakṣatras. The people learned in the Śāstras say that these devatā names are to be substituted for their own names in the saṅkalpa of yāga of the Yajamāna (the person on whose behalf the Yajña is performed).

Note: According to the Vedāṅga Jyotiṣa, the names are to be christened according to the Devatās (Vedic names) of the Nakṣatras and not on the basis of Nakṣatras.

उग्राण्यार्द्रा च चित्रा च विशाखा श्रवणोऽश्वयुक् ।

क्रूराणि तु मघाः स्वातिर्ज्येष्ठा मूलं यमस्य यत् ॥36॥

ugrāṇyārdrā cha chitrā cha viśākhā śravaṇo'śvayuk;
krūrāṇi tu maghāḥ svātirjyeṣṭhā mūlaṁ yamasya yat.

Ārdrā, Chitrā, Viśākhā, Śravaṇa and Aśvinī are ugra (fierce/sharp) Nakṣatras. Maghā, Svāti, Jyeṣṭhā, Mūla and Bharaṇi are Krura (cruel) Nakṣatras.

द्व्यूनं द्विषष्टिविभागेन दिनं सौराच्च पार्वणम् ।

यत्कृतावुपजायेते मध्येऽन्ते चाधिमासकौ ॥37॥

dvyūnaṁ dviṣaṣṭivibhāgena dinaṁ saurāchcha pārvaṇam;
yatkṛtāvupajāyete madhye'nte chā'dhimāsakau.

The Solar day is two cycle less than the 62 cycles of civil day. Due to this reason civil two extra months are produced in the middle of 5-year Yuga, i.e. Śuchi (Āṣāḍha) month and at the end of 5-year Yuga, i.e. Sahasya (Pauṣa) month.

In a half-yuga, there are 900 solar days, 915 civil days and 930 lunar days. So in half yuga one extra lunar month is produced in the middle of yuga. Similarly in the next half yuga also one more extra lunar month is produced as compared with the solar month. As such in the middle and at the end of a 5-year Yuga one extra lunar month, called adhika māsa is added.

कला दश सविंशाः स्यान् नाडिका ते मुहूर्तकः ।
द्यु त्रिंशत् तत् फलाना तु षड्सती त्र्यधिका भवेत् ॥38॥
kalā daśa saviṁśāḥ syān nāḍikā te muhūrtakaḥ;
dyu trimśat tat phalānā tu ṣaṭsatī trayadhikā bhavet.

One Nāḍikā is 10 plus one twentieth kalās (10+1/20, i.e. 10.05 kalās). Two Nāḍikās make one Muhurta. 30 Muhurtas (60 Nāḍikās) make one day which is equal to 603 kalās.

ससप्तकं भयुक् सोमः सूर्यो द्यूनि त्रयोदश।
नवमानि च पञ्चाऽह्नः काष्ठा पञ्चाक्षरा भवेत् ॥39॥
sasaptakaṁ bhayuk somaḥ sūryo dyūni trayodaśa;
navamāni cha pañchā'hnaḥ kāṣṭhā pañchākṣarā bhavet.

The Moon takes one day plus 7 kalās (i.e. 603+7=610 kalās) to move through one Nakṣatra. The Sun covers one Nakṣatra in 13 days and 9 parts of 5 days, i.e. 13+5/9 days, i.e. 13.5 days. The time required to pronounce 5 (akaṣaras) syllables makes one kāṣṭhā.

Note: According to Vāyupurāṇa (1.50.176) also, the time of sojourn of the Moon in one Nakṣatra is equal to a day (603 kalās) plus 7 kalās, i.e. 610 kalās.

The verse goes like this:

अहोरात्रं कलाश्चैव सप्त सोमः समश्नते ।

ahorātram kalāśchaiva sapta somaḥ samaśnate.

Length of Day in a Parva

यदुत्तरस्यायनतो गतं स्याच्छेषं तथा दक्षिणतोऽयनस्य ।
तदेकषष्ट्या द्विगुणं विभक्तं सद्वादशं स्याद् दिवसप्रमाणम् ॥40 ॥

yaduttarasyāyanato gatam syāchachheṣam tathā
dakṣiṇato'yanasya;
tadekaṣaṣṭyā dviguṇam vibhaktam sadvādaśam syād
divasapramāṇam.

The length of a given day (excluding night) is equal to the number of days that have elapsed from the beginning of the Uttarāyaṇa or which are required to complete the Dakṣiṇāyana multiplied by 2 and divided by 61, plus 12 (muhurtas).

Explanation: A day consists of 30 muhurtas. The Night in the Vedāṅga Jyotiṣa (śloka 8) was of 18 muhurtas on the day of Uttarāyaṇa; as such the day was 12 muhurtas long, the total difference being 6 muhurtas between the shortest day and longest night on Uttarāyaṇa. The year roughly consists of 366 days; therefore after a lapse of half-year, i.e. 183 days, the length of days and nights are in reverse position. The average rate of change in length of the day was therefore 6/183, or 2/61 or .0327868852 muhurtas per day. So the maximum and minimum length of a day would be 18 and 12 muhurtas respectively which has a ratio of 3:2.

Example: Finding out the length of the 13th day after Uttarāyaṇa.

Length of 13th day = 13x2 = 26÷61 =.4262295082+12 = 12.4262295082 muhurtas or mathematically we can say-

The Length of day= 12+ T (tithi) x (2/61)

यदर्धं दिनभागानां सदा पर्वणिपर्वणि ।

ऋतुशेषं तु तद् विद्यात् संख्याय सह पर्वणाम् ॥41॥

vadardhaṁ dinabhāgānāṁ sadā parvaniparvani;
ṛtuśeṣaṁ tu tad vidyāt saṁkhyāya saha parvaṇām. ॥41॥

For each parva we have half a portion of tithi. Note that its sum is Ṛtuśeṣa (tithis required to pass to complete solar ṛtu) along with a given number of parvas.

Example: To know the ṛtuśeṣa after 12 parvas, divide 12 by 2 and we get 6 ṛtuśeṣa along with the 12 parvas. This means 6 tithis have to pass to complete the Solar ṛtu.

Explanation: Solar parva (half month) is equal to 30.5÷2=15.25 days and Lunar parva (synodic half month) is equal to 29.5÷2=14.75 days. So there is a difference of .5 days which is called Ṛtuśeṣa per parva. After 12 parvas this difference (Ṛtuśeṣa) will increase to 12x.5=6 tithis which is needed to complete the solar ṛtu.

इत्युपायसमुद्देशो भूयोऽप्येनं प्रकल्पयेत् ।

ज्ञेयराशिंगताभ्यस्तं विभजेज् ज्ञातराशिना ॥42॥

ityupāyasamuddeśo bhūyo'pyenaṁ prakalpayet;
jñeyarāśiṁgatābhyastaṁ vibhajej jñātarāśinā.

On the basis of the above exposition, calendar may be made repeatedly for the coming years. Derive the unknown figures by the known figures by the process of repetition.

Valediction

इत्येवं मासवर्षाणां मुहूर्तोदयपर्वणाम् ।
दिनर्त्वयनमासानां व्याख्यानं लगधोऽब्रवीत् ॥43॥

ityevaṁ māsavarṣāṇāṁ muhūrtodayaparvaṇām;
dinartvayanamāsānāṁ vyākhyānaṁ lagadho'bravīt.

Thus did the sage Lagadha explain the synodic months, the year, the muhurtas, the risings, the parvas/pakṣas, the days, the ṛtus (seasons), ayanas, and solar months.

Benediction

सोमसूर्यस्तृचरितं विद्वान् वेदविदश्नुते ।
सोमसूर्यस्तृचरितं लोकं लोके च सन्ततिम्,
लोकं लोके च सन्ततिमिति ॥ 44 ॥

somasūryastṛcharitaṁ vidvān vedavidaśnute;
somasūryastṛcharitaṁ lokaṁ loke cha santatim;
lokaṁ loke cha santatimiti.

One learned in the Vedas who has also learnt the movement of the Moon, the Sun and the Stars, will attain all knowledge gained through the movements the Moon, the Sun and the Stars, besides having, in this world, a line of progeny.

आर्चज्योतिषम्

Ārchajyotiṣam

Introduction

पंचसंवत्सरमयं युगाध्यक्षं प्रजापतिम् ।
दिनर्त्वयनमासांगं प्रणम्य शिरसा शुचिः ॥ 1 ॥

paṁchasaṁvatsaramayaṁ yugādhyakṣaṁ prajāpatim;
dinartvayanamāsāṁgaṁ praṇamya śirasā śuchiḥ. ॥ 1 ॥

Having saluted with bent head to Prajāpati, the governor of five year yuga cycle involving various time-segments like the days, months, seasons and ayanas.

प्रणम्य शिरसा कालमभिवद्य सरस्वतीम् ।
कालज्ञानं प्रवक्ष्यामि लगधस्य महात्मनः ॥ 2 ॥

praṇamya śirasā kālamabhivadya sarasvatīm;
kālajñānaṁ pravakṣayāmi lagadhasya mahātmanaḥ.

Having slauted Time with a bent head, as also Sarasvatī, I explain the knowledge of Time as enunciated by sage Lagadha.

ज्योतिषामयनं पुण्यं प्रवक्ष्याम्यनुपूर्वशः ।
विप्राणां सम्मतं लोके यज्ञकालार्थ सिद्धये ॥ 3 ॥

jyotiṣāmayanaṁ puṇyaṁ pravakṣayāmyanupūrvaśaḥ;
viprāṇāṁ sammataṁ loke yajñakālārtha siddhaye.

I shall narrate in sequence the movement of heavenly bodies as accepted by the towering scholars of this field for the purpose of determining the time of yajñas.

Calculation of Parvarāśī

निरेकं द्वादशाभ्यस्तं द्विगुणं गतसंज्ञिकम् ।

षष्ट्या षष्ट्या युतं द्वाभ्यां पर्वणां राशिरुच्यते ॥ 4 ॥

nirekaṁ dvādaśābhyastaṁ dviguṇaṁ gatasaṁjñikam;
ṣaṣṭyā ṣaṣṭyā yutaṁ dvābhyāṁ parvaṇāṁ rāśiruchyate.

Take the given number of the year in the yuga, subtract this by 1, multiply by 12, again multiply by 2, and add the expired parvas of the given year; if the parva number is more than 60, add 2 for every 60 parvas, and the number obtained is the parva-rāśī (i.e. the total number of parvas expired at the time of calculation).

Note: In each year there are 12 synodic months, each month has 2 parvas. After 30 months, an intercalary month is added to complete the half yuga. That is why two parvas are added for each expired 60 parvas. Thus we get the total, there are 62 synodic months of 124 parvas in the yuga.

For example, to find the parva-rāśī (total parvas) in the 4th year (Anuvatsara) at end of the, Kārtika, Kṛṣṇa Aṣṭamī, Anuvatsara being the 4th year of the yuga, so (4-1), we get 3 years, after multiplying 3 by 12 (to convert years into months) we get, 36 and again multiplying 36 by 2 (to convert months into parvas/pakṣas), we get 72 parvas. The present year is Kārtika which is the 10th year. So the years elapsed since Kārtika so far are (10-1) 9. The parvas of 9 expired years will be 9x2=18. Total of no. of parvas are 72+18= 90. The n. of parvas are more than 60, so by adding 2 to 90, we get 92 expired parvas for the 10th Kārtika month of the Anuvatsara (or 4th year of 5-

year yuga).

Commencement of the Five-year Yuga

स्वराक्रमेते सोमार्कौ यदा साकं सवासवौ ।

स्यात्तदादियुगं माघस्तपः शुक्लोऽयनं ह्युदक् ॥ 5 ॥

svarākramete somārkau yadā sākaṁ savāsavau;
syāttadādiyugaṁ māghastapaḥ śuklo'yanaṁ hyudak.

When the Sun and the Moon rise up in Dhaniṣṭhā Nakṣatra (on an Amāvasyā day), at that time, on the first day of bright half of synodic lunar month Māgha, solar seasonal month Tapas and Uttarāyaṇa; commences the five year yuga-cycle.

The names of five Saṁvatsaras are as follows: Saṁvatsara, Parivatsara, Idāvatsara, Idvatsara and Vatsara.

The above statement shows that by the time of Yājuṣa Jyotiṣa, winter solstice was taking place in Dhaniṣṭhā Nakṣatra.

Note: One sidereal year consists of 365.25 days and one revolution of the Moon around the Earth with respect to a Nakṣatra takes place in 27.25 days. If we divide 365.25 by 27.25 we get 13.4036697248. This shows that the Moon and the Sun meet together on a year 13.4036697248 times. If we multiply the same by 5= we get 67.01. That means that the Moon and the Sun meet 67th time exactly in Dhanistha Nakstra (293.20^0-306.40^0) with a 1^0 difference in five years. Here it may be pointed out that the Sun and the Moon meet at the same point on the day of Amāvasyā.

प्रपद्यते श्रविष्ठादौ सूर्याचन्द्रमसावुदक् ।

सार्पार्धे दक्षिणार्कस्तु माघश्रावणयोः सदा ॥ 6 ॥

prapadyate śraviṣṭhādau sūryāchandramasāvudak;
sārpārdhe dakṣiṇārkastu māghaśrāvaṇayoḥ sadā.

When situated at the beginning of Dhaniṣṭhā Nakṣatra, the Sun and the Moon begin to move northward. When they reach the midpoint of Āśleṣā Nakṣatra, the sun begins to move southward. Thus Uttarāyaṇa begins in Māgha month and Dakṣiṇāyana begins in Śrāvaṇa month.

Note: Interestingly, Varāhamihira quotes the above statement of Vedāṅga Jyotiṣa while pointing out the difference of time between him and Vedāṅga Jyotiṣa. Varāhamihira states:

धर्मवृद्धिरपां प्रस्थः क्षपाह्रास उदग्गतौ ।
दक्षिणे तौ विपर्यासः षण्मुहूर्त्ययनेन तु ॥ 7 ॥

dharmavṛddhirapāṁ prasthaḥ kṣapāhrāsa udaggatau;
dakṣiṇe tau viparyāsaḥ ṣaṇmuhūrtyayanena tu.

During the northward course of the sun (Uttarāyaṇa), the increase of day-time and decrease in night-time per day is equal to the time taken by one prastha[6] (sher or litre) of water to enter the Nāḍī-yantra (a tubular instrument made for this purpose). The vice-versa happens during the southward course of the Sun. There is a total increase or decrease of time equal to 6

---

[6] Prastha (प्रस्थ) is a Sanskrit unit of weight corresponding to "400 grams" (or, 8 *palas*). It is commonly used in *Rasaśāstra* literature (Medicinal Alchemy) such as the *Rasaprakāśasudhākara* or the *Rasaratna-samuccaya*. Prastha is a weight-unit often used in various Ayurvedic recipes and Alchemical preparations

muhurtas.

Note: One muhurta is equal to around 48 minutes. So, 6 muhurtas are equal to (48x6=288 minutes=4 hours 48 minutes tentatively).

द्विगुणं सप्तमं चाहुरयनाद्यं त्रयोदश ।
चतुर्थं दशमं चैव द्विर्युग्मं बहुलेप्यृतौ ॥8॥

dviguṇaṁ saptamaṁ chāhurayanādyaṁ trayodaśa;
chaturthaṁ daśamaṁ chaiva dviryugmaṁ bahulepyṛtau.

Years can be divided into five classes according to the beginnings of ayanas (semester of solar years). First ayana start with 1st tithi, second with 7th tithi, the third will 13th tithi –all belonging to Śukla Pakṣa (bright half). The fourth ayana starts with 4th and 5th with 10th tithi-both belonging to Kṛṣṇa Pakṣa.

Note: The five-year yuga-cycle has 10 ayanas. So, the above calculation is for the first five ayanas to be repeated in the same order.

Explanation: The solar month consists of 30.5 days and the Synodic month consists of 29.5 days, so there is a difference of 1 tithi. As such a synodic month has 1 more tithi as compared to a solar month. In the ayana there are six synodic months, consequently, it has six tithis more. As such every 7th tithi comes as the beginning of ayana.

Note: There are 366 days in a year, so there will be 183 days in an ayana.

If the first ayana starts with the 1st tithi of Śukla Pakṣa; the second solar ayana will start 183 days after the first ayana, i.e. on the 184th day with the 7th tithi Śukla Pakṣa;

the third ayana will start 183 days after the second ayana, i.e. on 367th day with 13th tithi of Śukla Pakṣa; fourth will start 183 days after the third, i.e. on 551st day with 4th tithi of Kṛṣṇa Pakṣa; and finally the fifth ayana will start 183 days after the fourth, i.e. on 734th day with 10th tithi of Kṛṣṇa Pakṣa. The same cycle will repeat for another set of 5 ayanas (solstices). For example:

The 916th day will start from 1st tithi of Śukla Pakṣa

The 1099th day will start from 7th tithi Śukla Pakṣa

The 1282nd day will start from 13th tithi of Śukla Pakṣa

The 1466th day will start from 4th tithi of Kṛṣṇa Pakṣa

The 1649th day will start from 10th tithi of Kṛṣṇa Pakṣa

This can be verified from Vedāṅga Calendar revived and published by the present author.

Ayanas of five-year yuga cycle will commence with the following tithis.

Sr. No.	Name of the Year	1st Solstice (Winter Solstice)	2nd Solstice (Summer)
1	Saṁvatsara	Māgha S-1	Śrāvaṇa S-7
2	Parivatsara	Māgha S-13	Śrāvaṇa K-4
3	Idāvatsara	Māgha K-10	Śrāvaṇa S-1
4	Idvatsara	Māgha S-7	Śrāvaṇa S-13
5	Vatsara	Māgha K-4	Śrāvaṇa K-10

Nakṣatras at the Beginning of the Ayanas

वसुस्त्वष्टा भवोऽजश्च मित्रः सर्पोऽश्विनौ जलम् ।

धाता कश्चायनाद्याः स्युरर्धपंचमभस्त्वृतुः ॥9॥

vasustvaṣṭā bhavo'jaścha mitraḥ sarpo'śvinau jalam;

dhātā kaśchāyanādyāḥ syurardhapaṁchamabhastvṛtuḥ.

The 10 ayanas (solstices) of the five-year yuga cycle begin with one of the following nakṣatras:

The 10 ayanas (solstices) of the five-year yuga cycle begin with one of the following nakṣatras:

1. First ayana (1st day) with Vasu (Dhaniṣṭhā nakṣtra).
2. Second (184th day) with Tvaṣṭā (Chitrā nakṣtra).
3. Third (367th day) with Bhava (Ārdrā nakṣtra).
4. Fourth (550th day) with Aja (Pūrvabhādrapada nakṣtra).
5. Fifth (733rd day) with Mitra (Anurādhā nakṣtra).
6. Sixth (916th day) with Sarpa (Āśleṣā nakṣtra).
7. Seventh (1099th day) with Aśvinau (Aśvinī nakṣtra).
8. Eighth (1282nd day) with Jala (Pūrvāṣāḍhā nakṣtra).
9. Ninth (1465th day) with Dhātā (Uttaraphālgunī nakṣtra).
10. Tenth (1648th day) with Ka (Rohiṇī nakṣtra).

Explanation: The Moon moves through 1809 nakṣatras in the 10 ayanas of the yuga. So, it moves through about 1809÷10=181 nakṣatras in one ayana. This amounts to 6 sidereal revolutions and 19 nakṣatras. This would make total of 1810 nakṣatras in a yuga, so we have to reduce one nakṣatra at some stage. Lagadha does so at the end of the first ayana. At the end of the first ayana or 12th parva. At the end of the first ayana (12th parva) he drops Puṣya after Punarvasu and uses Āśleṣā, thus making 1809 nakṣatras.

Solar and Lunar Parva-bhāṁśas

(Nakṣatra-parts traversed by the Sun and the Moon at the

end of a Particular Parva)

भांशाः स्युरष्टकाः कार्याः पक्षा द्वादशकोद्गताः ।
एकादशगुणश्चोनः शुक्लेऽर्धं चैन्दवा यदि ॥10॥

bhāṁśāḥ syuraṣṭakāḥ kāryāḥ pakṣā dvādaśakodgatāḥ;
ēkādaśaguṇaśchonaḥ śukle'rdhaṁ chaindavā yadi.

In a group of 12 pakṣas/parvas, 8 bhāṁśa arise. For the remainder (less than 12 parvas/pakṣas) 11 bhāṁśas (per parva/pakṣa) arise. When you calculate lunar bhāṁśa for a parvan ending in full moon-day, add bhāṁśas equal to a half Nakṣatra (62 bhāṁśas).

Mathematically we can say that to find out the Parva-bhaṁśas (Nakṣatra-parts traversed by the Sun or the Moon at the end of a particular parva), divide the given parva by 12. Multiply the quotient by 8 and the remainder by 11 and add both. If it is Śukla pakṣa (bright half) and we want to know the bhāṁśas (Nakṣatra-parts) traversed by the Moon, then add half of the bhāṁśa (total 124 parvas), i.e. 62.

Example: Suppose we want to find out the bhāṁśa (Nakṣatra-parts) traversed by the Moon and the Sun at the end of the 73rd parva, as per instructions:

Divide 73 by 12 = 6 quotient, 1 remainder

Multiply 6 by 8 = 48 and multiply 1 by 11 = 11

Add both- 48 + 11 = 59.

So, we can say that at the end of the 73rd parva, the Sun will traverse 59 Nakṣatra-aṁśas (Nakṣatra-parts).

Since the parva is a bright fortnight, Nakṣatra-aṁśas (Nakṣatra-parts) to be traversed by the Moon will be

59+62=121.

Note:

1. If the number of Bhāṁśa (Nakṣatra-parts) crosses 124, then subtract 124 and the remaining parts will be considered as traversed by the Sun or the Moon as the case may be.

2. If we want to find out the name of the Nakṣatra of the Sun at the end of the 73rd parva, divide the Bhāṁśa (Nakṣatra-part) of that parva by 27 and the remaining will be Nakṣatra of Jāvādi series. For example, divide 59 solar Bhāṁśas of 73rd Parva (Nakṣatra-parts traversed by the Sun at the end of 73rd parva) by 27, the remainder is 5 which shows that the parva is ending with 59 Nakṣatra-aṁśas (parts) of the 5th, Pūrvāṣāḍhā Nakṣatra of the Jāvādi series.

Similarly, if we want to find out the Nakṣatra of the Moon at the end of the 73rd parva, divide lunar bhāṁśas, i.e. 121 by 27 and the remainder 13 shows that the 73rd parva ends with 121 Nakṣatra-parts of the 13th Nakṣatra, Punarvasu of Jāvādi series.

The proof of the rule is as follows:

In the yuga of 5 years containing 124 parvas, the Sun traverses 5x27= 135 nakṣatras. Thus during each parva it traverses 135÷124= 1+11/124 (or 1.0887096774) nakṣatras. 11/124 means 11th of 124 parts. At the end of 12 parvas, it is 12 (1+11/124)= 12+(12x11)/124=12+ 132/124= 12+1+8/124= 13+8/124. 13 in 12 parvas means, it has completed 13 Nakṣatras and 8/124 means, for every 12 parvas it accumulates by 8. For the remainder, there are 11 (aṁśas) parts for each remainder, so the remainder is multiplied by 11 and added. At the new moon, the Moon is with the Sun and aṁśas (parts) are the same. However, at full moon, the Moon is opposite to the Sun, that is, 13 Nakṣatras and 62 aṁśas (parts) away. So, we add 62 aṁśas (parts) for the Moon.

Below are given Solar and Lunar Parva-bhāṁśas (Nakṣatra-parts traversed by the Sun and the Moon at the end of the Parva.

Parva No.	Solar Nakṣatras and its Añśa (Bhañśa) at the end of the Parva	Lunar Nakṣatras and its Āñśa (Bhāñśa) at the end of the Parva
Saṁvatsara 1	Śatabhiṣak 11°	73 Maghā
2	22 P. Bhādrapada	22 P. Bhādrapada
3	33 U. Bhādrapada	95 Uttaraphalgunī
4	44 Revati	44 Revati
5	55 Aśvinī	117 Chitrā
6	66 Bharaṇī	66 Bharaṇī
7	77 Kṛttikā	15 Anurādhā
8	88 Rohiṇī	88 Rohiṇī
9	99 Mṛgaśirā	37 Mūla
10	110 Ārdrā	110 Ārdrā
11	121 Punarvasu	59 Uttarāṣāḍhā
12	8 Āśleṣā	8 Āśleṣā
13	19 Maghā	81 Dhaniṣṭhā
14	30 P. Phalgunī	30 P. Phalgunī
15	41 U. Phalgunī	103 Pūrvabhādrapada
16	52 Hasta	52 Hasta

17	63 Chitrā	1 Aśvinī
18	74 Svāti	74 Svāti
19	85 Viśākhā	23 Kṛttikā
20	96 Anurādhā	96 Anurādhā
21	107 Jyeṣṭhā	45 Mṛgaśira
22	118 Mūla	118 Mūla
23	5 Uttrāṣāḍhā	67 Punarvasu
24	16 Śravaṇa	16 Śravaṇa
Parivatsara 25	27 Dhaniṣṭhā	89 Āśleṣā
26	38 Śatabhiṣak	38 Śatabhiṣak
27	49 P. Bhādrapada	111 P. Phalgunī
28	60 U. Bhādrapada	60 U. Bhādrapada
29	71 Revati	9 Chitrā
30	82 Aśvinī	82 Aśvinī
31	93 Bharaṇī	31 Viśakhā
32	104 Kṛttikā	104 Kṛttikā
33	115 Rohiṇī	53 Jyeṣṭhā
34	2 Ārdrā	126-124=2 Ārdrā
35	13 Punarvasu	75 Pūrvāṣāḍhā
36	24 Puṣya	24 Puṣya
37	35 Āśleṣā	97 Śravaṇa
38	46 Maghā	46 Maghā
39	57 P. Phalgunī	119 Śatabhiṣak
40	68 U. Phalgunī	68 U. Phalgunī
41	79 Hasta	17 Revatī
42	90 Chitrā	90 Chitrā
43	101 Svāti	39 Bharaṇī

44	112 Viśākhā	112 Viśākhā
45	123 Anurādhā	61 Rohiṇī
46	10 Mūla	10 Mūla
47	21 Pūrvāṣāḍhā	83 Ārdrā
48	32 Uttarāṣāḍhā	32 Uttarāṣāḍhā
Idāvatsara 49	43 Śravaṇa	105 Puṣya
50	54 Dhaniṣṭhā	54 Dhaniṣṭhā
51	65 Śatabhiṣak	127-124=3 P. Phalgunī
52	76 P. Bhādrapada	76 P. Bhādrapada
53	87 U. Bhadrapada	25 Hasta
54	98 Revati	98 Revati
55	109 Aśvanī	47 Svāti
56	120 Bharaṇī	120 Bharaṇī
57	7 Rohiṇi	69 Anurādhā
58	18 Mṛgaśirā	18 Mṛgaśirā
59	29 Ārdrā	91 Mūla
60	40 Punarvasū	40 Punarvasū
61	51 Puṣya	113 Uttarāṣāḍhā
62	62 Āśleṣā	62 Āśleṣā
63	73 Maghā	11 Śatabhiṣak
64	84 P. Phalgunī	84 P. Phalgunī
65	100 U.Phalgunī	33 U. Bhādrapada
66	106 Hasta	106 Hasta
67	122 Chitrā	55 Aśvinī
68	4 Viśākhā	128-124=4 Viśākhā
69	15 Anurādhā	77 Kṛttikā
70	26 Jyeṣṭhā	26 Jyeṣṭhā

71	42 Mūla	99 Mṛgaśirā
72	48 Pūrvāṣāḍhā	48 Pūrvāṣāḍhā
73	59 Uttarāṣāḍhā	121 Punarvasu
74	70 Śravaṇa	70 Śravaṇa
Anuvatsara 75	81 Dhaniṣṭhā	19 Maghā
76	92 Śatabhiṣak	92 Śatabhiṣak
77	103 P. Bhādrapada	41 U. Phalgunī
78	114 U. Bhādrapada	114 U. Bhādrapada
79	1 Aśvinī	63 Chitrā
80	12 Bharaṇī	12 Bharaṇī
81	23 Kṛttikā	85 Viśākhā
82	34 Rohiṇī	34 Rohiṇī
83	45 Mṛgaśirā	107 Jyeṣṭhā
84	56 Ārdrā	56 Ārdrā
85	67 Punarvasū	5 Uttarāṣāḍhā
86	78 Puṣya	78 Puṣya
87	89 Āśleṣā	27 Dhaniṣṭhā
88	100 Maghā	100 Maghā
89	111 P. Phalgunī	49 P. Bhādrapada
90	122 U. Phalgunī	122 U. Phalgunī
91	9 Chitra	71 Revatī
92	20 Svāti	20 Svāti
93	31 Viśākhā	93 Bharaṇī
94	42 Anurādhā	42 Anurādhā
95	53 Jyeṣṭhā	115 Rohiṇī
96	64 Mūla	64 Mūla
97	75 Pūrvāṣāḍha	13 Punarvasu

98	86 Uttarāṣāḍhā	86 Uttarāṣāḍhā
Idavastsara 99	121 Śravaṇa	35 Āśleṣā
100	108 Dhaniṣṭhā	108 Dhaniṣṭhā
101	19 Śatabhiṣak	57 P. Phalgunī
102	6 U. Bhādrapad	6 U. Bhādrapad
103	41 Revati	79 Hasta
104	28 Aśvinī	28 Aśvinī
105	63 Bharaṇī	101 Svāti
106	50 Kṛttikā	50 Kṛttikā
107	85 Rohiṇī	123 Anurādhā
108	18 Mṛgaśirā	18 Mṛgaśirā
109	83 Ārdrā	21 Pūrvāṣāḍha
110	94 Punarvasu	94 Punarvasu
111	105 Puṣya	43 Śravaṇa
112	116 Āśleṣā	116 Āśleṣā
113	3 P. Phalgunī	65 Śatabhiṣak
114	14 U. Phalgunī	14 U. Phalgunī
115	25 Hasta	87 U. Bhādrapada
116	36 Chitrā	36 Chitrā
117	47 Svāti	109 Aśvinī
118	58 Viśākhā	58 Viśākhā
119	69 Anurādhā	7 Rohiṇī
120	80 Jyeṣṭhā	80 Jyeṣṭhā
121	91 Mūla	29 Ārdrā
122	102 Pūrvāṣāḍhā	102 Pūrvāṣāḍhā
123	113 Uttarāṣāḍhā	51 Puṣya
124	124 Śravaṇa	124 Śravaṇa

Solar/Lunar Parva Bhādānakalā

(Solar/Lunar Nakṣatra Start Time in Kalās at the end of a Parva)

कार्या भांशाष्टकास्थाने कला एकान्नविंशतिः ।
उनस्थाने त्रिसप्ततिमुद्द्वपेदूनसम्मिताः ॥ 11 ॥

*kāryā bhāṁśāṣṭakāsthāne kalā ēkānnaviṁśatiḥ;
unasthāne trisaptatimudvavapedūnasammitāḥ.*

To know the bhādānakalā of Moon or Sun (Lunar or Solar Nakṣatra start time in kalās) at the end of a given parva, substitute 19 kalās for every group of 8 bhaṁśas and take out from them 73 times the remainder bhāṁśas (when the remainder occurs).

Mathematically saying, divide the bhāṁśas at the end of parva by 8, multiply the quotient by 19 and take out the remainder multiplied by 73 and you will get Bhādānakalā of Moon at the end of a parva.

For example, if we want to find out the bhādānakalā of Moon at the end of the 6th parva. We have 66 bhāṁśas in the Moon's position at the end of the 6th parva. Divide 66÷8, we have quotient 8 and remainder 2 bhāṁśas. After applying the above rule, we have (8x19-2x73)= 6 bhadānakalās at the end of the 6th parva. That is the time of Moon's entry into the New Moon position at the end of the 6th parva.

Correction for the Sidereal day and the Five-Year Yuga Scheme

त्र्यंशो भशेषो दिवसांशभागश्चतुर्दशस्याऽप्युपनीय भिन्नम् ।

भार्धाेऽधिके चापि गते परोंऽशो द्वावुक्तमेकं नवकै र्खेद्यु ॥12 ॥

travaṁśo bhaśeṣo divasāṁśabhāgaśchaturdaśasyā'pyupanīya bhinnam |
bhārdhā'dhike chāpi gate paroṁ'śo dvāvuktamekaṁ navakai raverdyu ॥12 ॥

The nakṣatra in a sidereal circle is i/3ʳᵈ bhāṁśa more than the diurnal circle. That is, a sidereal circle has 124 bhāṁśas, but the diurnal circle has 124-1/3=124 2/3 or 123.6666666667 bhāṁśas. To bring the diurnal circle in unison with the sidereal circle, add the excess 9 bhāṁśas (or 1 tithi) arising out of the excess of 1/3ʳᵈ bhāṁśas mentioned above to the 14ᵗʰ tithi as the 15ᵗʰ tithi at the bhārdha (first half) parva (62ⁿᵈ parva) or the prāṁśa (end of the 2ⁿᵈ half) of the yuga, i.e. 124ᵗʰ parva (wherever the 15ᵗʰ tithi is dropped). Thus the above rule talks about the deletion of 29 tithis instead of 30 tithis as ruled in śloka 12.

Explanation: The Sidereal circle has 124x27=3348 bhāṁśas in one year and the diurnal circle will have 123.6666666667x27= 3339 bhāṁśa in a year. Thus 3348-3339=9 bhāṁśas will be in excess each year. 9 bhāṁśas are equal to 1 tithi. One tithi has 122 aṁśas. To bring the sidereal circle in unison with the diurnal circle this excess of 9 bhāṁśas or one tithi is to be added either when the sun crosses the first half number of parvas, i.e. at the end of 62ⁿᵈ parva and second-half number of parvas, i.e. 124ᵗʰ parva. Adding one tithi means retaining the 15ᵗʰ tithi if it is being dropped either at the end of 62ⁿᵈ parva or at the end of 124ᵗʰ parva as per rule prescribed in śloka no. 12 of Yājuṣa Jyotiṣa. We find in the calendar that the 15ᵗʰ tithi is dropped from both the parvas. So we can retain the 15th tithi either of 62nd parva or the 124th parva.

Explanation 1: One sidereal circle consisting of 367 days has 3348÷9=372 tithis and a diurnal circle consisting of 366 days has 3339÷9= 371 tithis. As such, there is a difference of 1 tithi. We are therefore required to add 3348-3339=9 bhāṁśas or 367-366=1 tithi.

Note: Here Chaturdaśī is mentioned for any addition or at the end of the parva because the 15th tithis (amāvasyās and pūrṇimās) are often dropped as kṣaya tithis.

Explanation 2: We are informed by the Vedāṅga Jyotiṣa that 5-year Yuga contains:

1. 60 solar months of 30.5 days each, which makes a Yuga of 60 x 30.5= 1830 solar days.

2. 61 sāvana months of 30 days each, which amounted to 61 x 30= 1830 sāvana days in a Yuga, requiring one adhikamāsa in 5 years for being in unison with solar months.

3. 62 synodic lunar months (tithimāsas) of 29.532 days each which makes a Yuga of 62 x 29.532=1830.984 or 1831 days, requiring two adhikamāsas in 5 years for being in unison with solar months.

4. 67 sidereal lunar months (nakṣatra months) of 27.32 days each, which amounted to 67 x 27.32=1830.44 days, requiring 7 adhikamāsās in 5 years for being in unison with solar months.

In other words, 1830 is the number of days in one

yuga represented a least common multiple (LCM) of solar, sāvana, synodic luna and sidereal lunar months. However in reality a yuga contained 1831 days. So, the length of the Yuga has to be increased by one day or 124 aṁśas in five years. By retaining one tithi at the end of the 62nd or 124th parva we may add 122 aṁśas. The rest of the two aṁśas may also be added at the end of the 124th parva. So the 15th tithi of 124th parva will have 124 aṁśas.

Hour Angle and Langna of a Nakṣatra at the End of a Parva

पक्षात्पंचदशाच्चोर्ध्वं तद्भुक्तमिति निर्दिशेत् ।

नवभिस्तूद्गतोंऽशः स्यादूनांशद्वयधिकेन तु ॥ 13 ॥

pakṣātpaṁchadaśāchchordhvaṁ tadbhuktamiti nirdiśet;
navabhistūdgatoṁ'śaḥ syādūnāṁśadvayadhikena tu.

To tell the Hour angle and Lagna at the end of the 15th tithi of any parva (end of parva), divide the number of parvas by 9. Each of the remainder should be multiplied by 2 parts less than 9 (i.e. 7) and added to the quotient. [If the quotient is odd, add half of the diurnal circle (that is 62 parts). If the Paulastya (Moon) is setting when the sun rises (i.e. of full moon parva), add another half (i.e. 62 parts)]

For example, if we want to know the hour angle and lagna (ascendant) of a Nakṣatra at the end of the 93rd parva, divide the 93rd parva by 9, and we get 10 as quotient and 3 remainder. Multiply remainder by 7, i.e. 3x7=21 and add to the quotient 10 = 31. 93 is full moon parva, so add 62 to 31= 93. As such we are able to know that by the end of the 93rd parva, the 93rd aṁśas of nakṣatra is the rising point.

If we want to know the name of nakṣatra, subtract from bhāṁśa of 93rd parva the multiples of 27, i.e. 93-(27X3)=12. It shows that at the end of the 93rd parva, 93 aṁśas of the 12th nakṣatra named Bharaṇī of the Jāvādi list is the rising point, lagna.

Note: Aṁśa of tithi on a particular day will be known as the hour angle of the Sun. By subtracting the hour angle of the Sun (tithi aṁśas) from the nakṣatra-aṁśas obtained, we get the hour angle of Dhaniṣṭhā.

For example, by subtracting the Sun's hour angle from obtained nakṣatra aṁśas, we get 93-62=31 as the hour angle of Dhaniṣṭhā. 31x27÷124=6.75, i.e. 6th nakṣatra counted from Dhaniṣṭha, i.e. Bharaṇi is again found as rising, lagna at the end of the 93rd parva.

जौ द्रा गः खे श्वे ही रो षा श्चिन्मूषकण्यः सूमाधाणः ।

रे मृ घाः स्वापोजः कृष्यो ह ज्येष्ठा इत्यृक्षा लिंगैः ॥ 14 ॥

jau drā gaḥ khe śve hī ro ṣā śchinmūṣakṇyaḥ sūmādhāṇaḥ;
re mṛ ghāḥ svāpojaḥ kṛṣyo ha jyeṣṭhā ityṛkṣā liṁgaiḥ.

Take the Nakṣatras represented symbolically in the Jāvādi list in order of 1,2,3 (which is different from the normal order of the Nakṣatras). So many aṁśas (parts) of that Nakṣatra have gone at the end of that parva for which that bhāṁśa has been found.

Explanation: In a year, the Moon goes ahead of the Sun by 27+10 nakṣatras or say 13.5 + 5 nakṣatras in an ayana (semester). The five nakṣatras difference in the Jāvādi list is attributed to this circumstance.

Nakṣatra Table with Symbols

No.	Symbol	Nakṣatra	No	Symbol	Nakṣatra
1	Jau	Āśvayujau	14	Mā	Aryamā (Uttaraphalgunī)
2	Dra	Ārdrā	15	Dhāḥ	Anurādhāḥ
3	Gaḥ	Bhagḥ (Pūrva Phalgunī)	16	Naḥ	Śravaṇaḥ
4	Khe	Viśākhe	17	Re	Revati
5	Śve	Viśvedevāḥ (Uttarāṣāḍhā)	18	Mṛ	Mṛgaśirāḥ
6	Hiḥ	Ahirbudhnyaḥ (U. Bhādrapada)	19	Ghaḥ	Maghaḥ
7	Ro	Rohiṇī	20	Svā	Svāti
8	Ṣā	Āśleṣā	21	Paḥ	Āpaḥ (Pūrvāṣāḍhā)
9	Cit	Chitrā	22	Jaḥ	Aja ekapāt (P. Bhādrapada)
10	Mū	Mūla	23	Kṛ	Kṛttikāḥ
11	Śa	Śatabhiṣak	24	Ṣyaḥ	Puṣyaḥ
12	Ṇyaḥ	Bharaṇyaḥ	25	Ha	Hastaḥ
13	Su	Punarvasu	26	Jye	Jyeṣṭhā
			27	Ṣṭhāḥ	Śraviṣṭhāḥ (Dhaniṣṭhā)

Note: In the Javādi list, Nakṣatras have been arranged from Aśvinī, each being the sixth from the normal arrangement of Nakṣatras.

The normal order of the Nakṣatras reckoned from Aśvinī is given below:

1. Aśvinī
2. Bharaṇi
3. Kṛttikā
4. Rohiṇi
5. Mṛgaśirā
6. Ārdrā
7. Punarvasu
8. Puṣya

9. Aśleṣā	10. Maghā
11. Pūrva Phalguni	12. Uttara Phalguni
13. Hasta	14. Chitrā
15. Svāti	16. Viśākhā
17. Anurādhā	18. Jyeṣṭhā
19. Mūla	20. Pūrvāṣāḍhā
21. Uttarāṣāḍhā	22. Śrāvaṇa
23. Dhaniṣṭhā	24. Śatabhiṣak
25. Pūrva Bhādrapada	26. Uttara Bhādrapada
27. Revati	

Order of the Nakṣatras reckoned from Dhaniṣṭhā (Śraviṣṭhā) Aśvinī is given below:

1. Dhaniṣṭhā	2. Śatabhiṣak
3. Pūrva Bhādrapada	4. Uttara Bhādrapada
5. Revati	6. Aśvinī
7. Bharaṇ	8. Kṛttikā
9. Rohiṇi	10. Mṛgaśira
11. Ārdrā	12. Punarvasu
13. Puṣya	14. Aśleṣā
15. Maghā	16. Pūrva Phalgunī
17. Uttara Phalgunī	18. Hasta
19. Chitrā	20. Svāti
21. Viśākhā	22. Anurādhā
23. Jyeṣṭhā	24. Mūla
25. Pūrvāṣāḍhā	26. Uttarāṣāḍhā
27. Śravaṇa	

Jāvādi List of Nakṣatras

जावाद्यंशैः समं विद्यात् पूर्वार्धे पर्व सूत्तरे ।
भादानं स्याच्चतुर्दश्यां काष्ठानां देविना कलाः ॥ 15 ॥

jāvādyaśaiḥ samaṁ vidyāt pūrvārdhe parva sūttare;
bhādānaṁ syāchchaturdaśyāṁ kāṣṭhānāṁ devinā kalāḥ.

Know that the Jāvādi list is balanced. The former letter in the list represents the lunar nakṣatra at the beginning of the first half. If the parva falls within the first half of the bhāṁśa, i.e. if the bhāṁśa is 62 or less at parva, the beginning of the Nakṣatra (bhādānam) will fall in the parva tithi (fortnight-ending), i.e. 15th tithi itself. If the Nakṣatra part is greater than the parts of the day at which the parva falls, the beginning of the Nakṣatra falls on the Chaturdaśī tithi day.

Time Measures

कला दश सविंशा स्याद् द्वे मुहुर्तस्य नाडिके ।
द्युस्त्रिंशत् तत्कलानां तु षड्सती त्र्यधिका भवेत् ॥16 ॥

kalā daśa saviṁśā syād dve muhurtasya nāḍike;
dyustrimśaṁt tatkalānāṁ tu ṣaṭsatī trayadhikā bhavet.

One Nāḍikā is 10 plus one-twentieth kalās (10+1/20, i.e. 10.05 kalās). Two Nāḍikās make one Muhurta. 30 Muhurtas (60 Nāḍikās) make one day which is equal to 603 kalās.

नाडिके द्वे मुहुर्तस्तु पंचाशत्पलमाढकम् ।
आढाकात्कुम्भकोद्रोणः कुटपैर्वर्धते त्रिभिः ॥ 17 ॥

nāḍike dve muhurtastu pamchāśatpalamāḍhakam;
āḍhakātkumbhākodroṇaḥ kuṭapairvardhate tribhiḥ.

One Muhūrta is equal to two Nāḍikās. An Āḍhaka is equal to 50 palas. From āḍhaka, kumbhaka or droṇa increases by three kuṭapas (kuḍavas).

Note: Droṇa is also measured by a kumbha (pot). So, droṇa and kumbhaka are synonymous.

ससप्तकं भयुक् सोमः सूर्यो द्यूनि त्रयोदश ।
नवमानि च पंचाह्नः काष्ठाः पंचाक्षरा भवेत् ॥ 18 ॥

sasaptakaṁ bhayuk somaḥ sūryo dyūni trayodaśa ;
navamāni cha paṁchāhnaḥ kāṣṭhāḥ paṁchākṣarā bhavet.

When Moon moves through a Nakṣatra, it takes 7 kalās more than a day. When the Sun moves through a Nakṣatra, it takes 13 days plus the ninth part of 5, (i.e. .5). This can be represented as 13+5/9 or 13.5 days. Five akṣaras make one kāṣṭhā (Time required to pronounce one syllable is called 'akṣara').

Note: A day has 603 kalās, but the Moon takes 610 kalās which are 7 times more than a day.

Note: According to Vāyupurāṇa (1.50.176) also, the time of sojourn of the Moon in one Nakṣatra is equal to a day (603 kalās) plus 7 kalās, i.e. 610 kalās.

The verse goes like this:

अहोरात्रं कलाश्चैव सप्त सोमः समश्नते ।
ahorātraṁ kalāśchaiva sapta somaḥ samaśnate.

Total No. of Lagnas

श्रविष्ठाभ्यो गणाभ्यस्तात् प्राग्विलग्नान् विनिर्दिशेत् ।
स्तर्यान् मासान् षडभ्यस्तान् विद्याच्चान्द्रमसानृतून ॥ 19 ॥

śraviṣṭhābhyo gaṇābhyastān prāgvilagnān vinirdiśet;
staryān māsān ṣaḍabhyastān vidyāchchāndramasānṛtūna.

To find the total no. of lagnas or risings of Dhaniṣṭha

etc. nakṣatras on the eastern horizon in a yuga of 1835 days, multiply 1835x27, we get = 49545 total lagnas; and multiplying the sidereal revolutions of moon in a yuga, i.e. 67 by 6, we get the total number of lunar ṛtus, i.e. 67x6=402 ṛtus.

End Time of a Tithi in Aṁśas and Diurnal Position of the Sun

अतीतपर्वभागेभ्यः शोधयेद् द्विगुणां तिथिम् ।

तेषु मंडलभागेषु तिथिनिष्ठांगतो रविः ॥20॥

atītaparvabhāgebhyaḥ śodhayed dviguṇāṁ tithim;
teṣu maṁḍalabhāgeṣu tithiniṣṭhāṁgato raviḥ.

Multiply the given tithi by 2 and subtract it from the tithi aṁśa (time of the end of the tithi in aṁśas) ending the previous parva. This is the same as the position of the Sun in the Nāḍimaṇḍal (diurnal circle).

If we want to know the end time of the 8th tithi after the 93rd parva, subtract 8x2=16 from the tithi aṁśa (part of the day in aṁśas), 92 ending the previous (92nd) parva. The result is 92-16=76.

Tithi Bhādānikā/Bhādānakalā

(Tithi Nakṣatra End Time in Kalās)

याः पर्वाभादानकलास्तासु सप्तगुणां तिथिम् ।

पक्षिपेत्तत् समूहस्तु विद्यादादानिकाः कलाः ॥ 21 ॥

yāḥ parvābhādānakalāstāsu saptaguṇāṁ tithim;
pakṣipettat samūhastu vidyādādānikāḥ kalāḥ.

Add elapsed tithi multiplied by 7 to the bhādāna kalā (beginning of the Nakṣatra) of the elapsed parva, we get the Bhādānakalā (Nakṣatra start time) of tithis.

Explanation: The duration of the Moon in a Nakṣatra is 610 kalās and the duration of the Moon in a tithi is 603 kalās, so a Nakṣatra begins 7 kalās later daily as compared to the Bhādāna-kalā of a tithi. Hence the rule of adding 7 times the tithis, so that Nakṣatra may come in consonance with a Parvabhādāna -kalā.

Example: To find the Bhādāna-kalā of the fifth tithi of the 13th parva, multiply elapsed 4x7=28 and add it to the bhādānakalā of the elapsed parva 12, i.e. 19. 28+19=47.

Similarly, if we want to find out the 8th tithi of 93rd parva, multiply 7 elapsed tithis by 7=49. Add Bhādānakalā of the elapsed parva 92, 356 to it and we get the Bhādāna-kalā of 8th tithi of 93rd Nakṣatra as 49+356=405.

Length of a Day in a Parva

यदुत्तरस्यायनतो गतं स्याच्छेषं तथा दक्षिणतोऽयनस्य ।
तदेकषष्ट्याद्द्विगुणं विभक्तं सद्द्वादशं स्याद् दिवसप्रमाणम् ॥ 22 ॥

yaduttarasyāyanato gatam syāch chheṣam tathā
dakṣiṇato'yanasya;
tadekaṣaṣṭyādviguṇam vibhaktam sadvādaśam syād
divasapramāṇam.

The length of a given day (excluding night) is equal to the number of days that have elapsed from the beginning of the Uttarāyaṇa or which are required to complete the Dakṣiṇāyana multiplied by 2 and divided by 61, plus 12 (muhurtas).

Explanation: A day consists of 30 muhurtas. The Night in the Vedāṅga Jyotiṣa (śloka 8) was of 18 muhurtas on the day of Uttarāyaṇa; as such the day was 12 muhurtas long, the total difference being 6 muhurtas. The year roughly consists of 366 days; therefore after a lapse of half-year, i.e. 183 days, the length of days and nights are in reverse position. The average rate of change in length of the day was therefore 6/183, or 2/61 or .0327868852 muhurtas per day. The minimum length of a day was 12 muhurtas.

Example: Find out the length of the 13th day after Uttarāyaṇa.

Length of 13th day = 13x2 = 26÷61 =.4262295082+12 = 12.4262295082 muhurtas or mathematically we can say-

Length of day= 12+ T (tithi) x (2/61)

यदर्धं दिनभागानां सदा पर्वणि पर्वणि ।

ऋतुशेषं तु तद् विद्यात् संख्याय सह पर्वणाम् ॥ 23 ॥

yadardhaṁ dinabhāgānāṁ sadā parvaṇi parvaṇi;
ṛtuśeṣaṁ tu tad vidyāt saṁkhyāya saha parvaṇām.

For each parva we have half a portion of tithi. Note that its sum is Ṛtuśeṣa (tithis required to pass to complete solar ṛtu) along with the given number of parvas.

Example: To know the ṛtuśeṣa after 12 parvas, divide 12 by 2 and we get 6 ṛtuśeṣa along with the 12 parvas. This means 6 tithis have to pass to complete the Solar ṛtu.

Explanation: Solar parva (half month) is equal to 30.5÷2=15.25 days and Lunar parva (synodic half month) is equal to 29.5÷2=14.75 days. So there is a difference of .5 days. After 12 parvas this difference will increase to

12x.5=6 tithis which need to complete the solar ṛtu.

इत्युपायसमुद्देशो भूयोप्येनं प्रकल्पयेत् ।
ज्ञेयराशिं गताभ्यस्तान्विभजेज्ज्ञानराशिना ॥ 24 ॥

ityupāyasamuddeśo bhūyopyahnaḥ prakalpayet;
jñeyarāśiṁ gatābhyastaṁ vibhajejjñānarāśinā.

The above exposition can be used to formulate the calendar time and again we can derive the unknown figures by the known figures by the process of repetition (to prepare a calendar of a given year).

Vedic Names of Nakṣatras

अग्निः प्रजापतिः सोमो रुद्रोदितिबृहस्पती ।
सर्पाश्च पितरश्चैव भगश्चैवार्यमापि च ॥ 25 ॥

agniḥ prajāpatiḥ somo rudroditibṛhaspatī;
sarpāścha pitaraśchaiva bhagaśchaivāryamāpi cha.

सविता त्वष्टाथ वायुश्चेन्द्राग्री मित्र एव च ।
इन्द्रो निर्ऋतिरापो वै विश्वेदेवास्तथैव च ॥26॥

savitā tvaṣṭātha vāyuśchendrāgnī mitra ēva cha;
indro nirṛtirāpo vai viśvedevāstathaiva cha.

विष्णुर्वसवो वरुणऽजेकपात् तथैव च ।
अहिर्बुध्न्यस्तथा पूषा अश्विनौ यम एव च ॥ 27 ॥

viṣṇurvasavo varuṇū'jekapāt tathaiva cha;
ahirbudhnyastathā pūṣā aśvinau yama ēva cha.

The Devatas (Vedic names) of various Nakṣatras are as under:

Sr. No.	Devata (Vedic Name)	Name of Nakṣatra

1	Agni	Kṛttikā
2	Prajāpati (Brahmā)	Rohiṇī
3	Soma	Mṛgaśirā
4	Rudra	Ārdrā
5	Aditi	Punarvasu
6	Bṛhaspati	Tiṣya
7	Sarpa	Āśleṣā
8	Pitara	Maghā
9	Bhaga	Purva Phālgunī
10	Aryamā (Dhātā)	Uttara Phalgunī
11	Savitā	Hasta
12	Tvaṣṭā	Chitrā
13	Vāyu	Svāti
14	Indrāgnī	Viśākhā
15	Mitra	Anurādhā
16	Indra	Jyeṣṭhā
17	Nirṛti	Mūla
18	Āpaḥ	Pūrvāṣāḍhā
19	Viśvedeva	Uttarāṣāḍhā
20	Viṣṇu	Śravaṇa
21	Vasu	Dhaniṣṭhā
22	Varuṇa	Śatabhiṣak
23	Aja Ekapāt	Pūrvabhādrapada
24	Ahirbudhnya	Uttara Bhādrapada
25	Pūṣā	Revatī
26	Aśvinīkumāra	Aśvinī
27	Yama	Bharaṇī or Apabharaṇī

Naming Based upon Vedic Names of the

Nakṣatras

नक्षत्रदेवता एता एताभिर्यज्ञकर्मणि ।
यजमानस्य शास्त्रज्ञैर्नाम नक्षत्रजं स्मृतम् ॥ 28 ॥

nakṣatradevatā ētā ētābhiryajñakarmaṇi;
yajamānasya śāstrajñairnāma nakṣatrajaṁ smṛtam.

These are the Devatās (Vedic names) of the Nakṣatras. The people learned in the Śāstras say that these devatā names are to be substituted for their own names in the saṅkalpa of yāga of the Yajamāna (the person on whose behalf the Yajña is performed).

Note: According to the Vedāṅga Jyotiṣa, the names are to be christened according to the Devatās (Vedic names) of the Nakṣatras and not on the basis of Nakṣatras.

Valediction

इत्येवं मासवर्षाणां मुहुर्तोदयपर्वणाम् ।
दिनर्त्वयनमासांगं व्याख्यानं लगधोऽब्रवीत् ॥ 29 ॥

ityevaṁ māsavarṣāṇāṁ muhurtodayaparvaṇām;
dinartvayanamāsāṁgaṁ vyākhyānaṁ lagadho'bravīt.

Thus the sage Lagadha described the months, the years, the muhūrtas, the risings of nakṣatras, the parvas, the days, the seasons and the ayanas.

Benediction

सोमसूर्यस्तृचरितं लोकांल्लोके च सम्मतिम् ।
सोमसूर्यस्तृचरितं विद्वान् वेदविदश्नुते ॥ 30 ॥

somasūryastṛcharitaṁ vidvān vedavidaśnute;
somasūryastṛcharitaṁ lokaṁ loke cha sammatim.

The motions of the Moon, the Sun and the Constellations are known to the common people. These motions of the Moon, the Sun and the constellations are analysed by the Vedic scholars mathematically.

Day of Viṣuva (Vernal Equinox)

विषुवं तद्गुणं द्वाभ्यां रूपहीनं तु षड्गुणम् ।
यल्लब्धं तानि पर्वाणि तथार्धं सा तिथिर्भवेत् ॥ 31 ॥

viṣuvaṁ tadguṇaṁ dvābhyāṁ rūpahīnaṁ tu ṣaḍguṇam;
yallabdhaṁ tāni parvāṇi tathārdhaṁ sā tithirbhavet.

Take the number of viṣuva (equinox) and multiply by 2; subtract 1; and multiply the balance by 6. What is obtained is the number of elapsed parvas. Half of this is the tithi at the end of which Viṣuva (equinox) occurs.

Note: There are 10 equinoxes in a five-year yuga of Vedāṅga Jyotiṣa, every year there are two equinoxes. The vernal equinox will be numbered one and Autumn equinox will be numbered two in the first year, the second year equinoxes will be 3rd and 4th, so on up to 10th.

Example: To know the tithi of the 3rd equinox, multiply 3 by 2=6. Subtract one 6-1=5, and multiply the balance 5 by 6=30. 30 will represent here elapsed parvas. Half of 30, i.e. 15 will be the 3rd Viṣuvān tithi or the day of the 3rd equinox (vernal equinox).

Nominal Definition of a Yuga

माघशुक्लप्रवृत्तस्य पौषकृष्णसमापिनः ।
युगस्य पंचमस्येह कालज्ञानं निबोधत ॥ 32 ॥

māghaśuklaprapannasya pauṣakṛṣṇasamāpinaḥ;

yugasya paṁchavarṣasya kālajñānaṁ prachakṣate.

Now the details will be given of five years of the yuga that starts with the bright fortnight of the month Māgha and ends with the dark fortnight of the month of Pauṣa. According to the Vedāṅga Jyotiṣa, the months are Amānta as per South Indian tradition. They start from Māgha Śukla Pratipadā and ends with Pauṣa Amāvasyā.

सौर संक्रान्ति	मास का नामकरण
मकर	माघ/ तप
कुम्भ	फाल्गुन/ तपस्य
मीन	चैत्र/मधु
मेष	वैशाख/माधव
वृषभ	ज्येष्ठ/शुक्र
मिथुन	अषाढ/शुचि
कर्क	श्रावण/नभ
सिंह	भाद्रपद/नभस्य
कन्या	आश्विन/इष
तुला	कार्तिक/ऊर्ज
वृश्चिक	मार्गशीर्ष/सह
धनु	पौष/सहस्य

Solar Saṁkrānti	Name of the Month
Makara	Māgha/ Tapa
Kumbha	Phālguna/Tapasya
Mīna	Chaitra/Madhu
Meṣa	Vaiśākha/Mādhava
Vṛṣabha	Jyeṣṭha/Śukra
Mithuna	Āṣāḍha/Śuchi

Karka	Śrāvaṇa/Nabha
Siṁha	bhādrapada/nabhasya
Kanyā	Āśvina/Iṣa
Tulā	Kārtika/Ūrja
Vṛschika	Mārgaśīrṣa/Saha
Dhanu	Pauṣa/Sahasya

Tithis of Viṣuvas

तृतीयां नवमीं चैव पौर्णमासीमथासिते ।

षष्ठि च विषुवान् प्रोक्तो द्वादशीं च समं भवेत् ॥33॥

tṛtīyāṁ navamīṁ chaiva paurṇamāsīmathāsite;
ṣaṣṭhi cha viṣuvān prokto dvādaśīṁ cha samaṁ bhavet.

The Viṣuva (Vernal Equinox) occurs in the Śukla Pakṣa at 3rd, 9th and 15th tithis and Kṛṣṇa Pakṣa at 6th and 12th tithis. This is repeated once again.

Note: In a Yuga of 5 years, there are 10 ayanas. Viṣuva (Vernal Equinox) occurs in the middle of each ayana. The interval between the Viṣuva is 124/10 parvas=12.4, i.e 12 parvas and (.4x15 tithis=) 6 tithis. So each viṣuva will occur at the interval of 6 tithis.

Sign of the Commencement of a New Yuga

चतुर्दशीमुपवसथस्तथा भवेद्यथोदितो दिवसमुपैति चन्द्रमाः ।

माघशुक्लाह्निको भुङ्क्ते श्रविष्ठायां च वार्षिकीम् ॥ 34 ॥

chaturdaśīmupavasathastathā bhavedyathodito
divasamupaiti chandramāḥ;
maghaśuklāhniko bhuṅkte śraviṣṭhāyāṁ cha vārṣikīm.

The 14th tithi of Kṛṣṇa Pakṣa occurring at the end of

the yuga on which the Moon rises just after the rise of the Sun (that is when the Moon is found just above the rising Sun) is the day of Upavasatha. Thereafter, the new yuga begins on the next day with Śukla Pakṣa with the Sun and the Moon in Dhaniṣṭhā Nakṣatra.

Note: The 14th and 15th tithis of the Kṛṣṇa Pakṣa of the ending yuga are the same, as the 14th tithis is deleted as per rules. So, the new Yuga is said to have begun just after the end of the 14th tithi which is the same as the 15th. For details, refer to our Vedāṅga Calendar.

Importance of Mathematical Astronomy

यथा शिखा मयूराणां नागानां मणयो यथा ।
तद्वद्वेदांगशास्त्राणां ज्यौतिषं मूर्धनि स्थितम् ॥ 35 ॥

yathā śikhā mayūrāṇāṁ nāgānāṁ maṇayo yathā;
tadvadvedāṁgaśāstrāṇāṁ jyautiṣaṁ mūrdhāni sthitam.

Like the crest of peacock and jewels of serpents stands atop their head, so does the mathematical astronomy stand atop all Vedic Śāstras.

वेद हि यज्ञार्थमभिप्रवृत्ताः कालानुपूर्व्या विहिताश्च यज्ञाः ।
तस्मादिदं कालविधानशास्त्रं यो ज्योतिषं वेद स वेद यज्ञान् ॥ 36 ॥

veda hi yajñārthamabhipravṛttāḥ kālānupūrvyā vihitāścha yajñāḥ;
tasmādidaṁ kālavidhānaśāstraṁ yo jyotiṣaṁ veda sa veda yajñān.

Vedas have indeed been revealed by the Brahman for the sake of understanding yajña (the process of creation). The process of creation involves the sequence of time. Therefore, the one who knows Astronomy, the Śāstra of calculating time, understands the process of creation.

इत्यार्चज्योतिषं समाप्तम् ॥ *iti ārchajyotiṣaṁ samāptam.*
Here ends the Ṛg-jyotiṣa.